Assessing Needs for Civilian STEM Talent in the Department of the Air Force

Navigating the Complexities of Workforce Demand and Supply

SEAN ROBSON, LISA M. HARRINGTON, KELLY ATKINSON, ALVIN MOON,
NATHAN THOMPSON, JONAH KUSHNER, GRACE FALGOUST,
ALICE NGUYEN, BARBARA BICKSLER, JOANNA ZAKZEWSKI

Prepared for the Department of the Air Force
Approved for public release; distribution is unlimited.

 PROJECT AIR FORCE

For more information on this publication, visit **www.rand.org/t/RRA3170-1**.

About RAND

RAND is a research organization that develops solutions to public policy challenges to help make communities throughout the world safer and more secure, healthier and more prosperous. RAND is nonprofit, nonpartisan, and committed to the public interest. To learn more about RAND, visit www.rand.org.

Research Integrity

Our mission to help improve policy and decisionmaking through research and analysis is enabled through our core values of quality and objectivity and our unwavering commitment to the highest level of integrity and ethical behavior. To help ensure our research and analysis are rigorous, objective, and nonpartisan, we subject our research publications to a robust and exacting quality-assurance process; avoid both the appearance and reality of financial and other conflicts of interest through staff training, project screening, and a policy of mandatory disclosure; and pursue transparency in our research engagements through our commitment to the open publication of our research findings and recommendations, disclosure of the source of funding of published research, and policies to ensure intellectual independence. For more information, visit www.rand.org/about/research-integrity.

RAND's publications do not necessarily reflect the opinions of its research clients and sponsors.

Published by the RAND Corporation, Santa Monica, Calif.
© 2025 RAND Corporation
RAND® is a registered trademark.

Library of Congress Cataloging-in-Publication Data is available for this publication.

ISBN: 978-1-9774-1503-5

Cover: Bradley Hicks/U.S. Air Force.

About This Report

This report addresses the Department of the Air Force's (DAF's) objective to enhance technical capabilities across its civilian workforce. Challenges in determining the DAF's science, technology, engineering, and mathematics (STEM) talent needs for its civilian workforce include tracking the supply of such talent and identifying potential gaps in technical competencies. This analysis centers on three case studies conducted with the following organizations: the Air Force Life Cycle Management Center, Digital Directorate (AFLCMC/HB) (Program Executive Office Digital and Enterprise Services [PEO Digital, now the Electronic Systems Directorate, or ESD[1]]); the Deputy Chief of Staff, Air Force Futures; and Headquarters, Pacific Air Forces (HQ PACAF). Key findings highlight decentralized information on technical talent demands, varied definitions of STEM, and barriers within the civilian personnel system. The report offers recommendations to support local organizations in identifying critical skills, improve data collection, and develop a DAF-wide competency framework in order to enhance workforce planning and addressing technical skill gaps.

The primary audiences for this report are policymakers and leaders within the DAF, particularly those involved in workforce planning and human capital management. Additionally, senior officials responsible for STEM talent development and organizational strategy will find the insights and recommendations valuable.

The research reported here was commissioned by the Air Force Chief Scientist and conducted within the Workforce, Development, and Health Program of RAND Project AIR FORCE (PAF) as part of a fiscal year (FY) 2024 project, "Science, Technology, Engineering and Math (STEM) Workforce Needs for the Development of Future Technology Required to Defeat China."

RAND Project AIR FORCE

RAND PAF, a division of RAND, is the DAF's federally funded research and development center (FFRDC) for studies and analyses, supporting both the U.S. Air Force (USAF) and the U.S. Space Force (USSF). PAF provides the DAF with independent analyses of policy alternatives affecting the development, employment, combat readiness, and support of current and future air, space, and cyber forces. Research is conducted in four programs: Strategy and Doctrine; Force Modernization and Employment; Resource Management; and Workforce, Development, and Health. The research reported here was prepared under contract FA7014-22-D-0001.

Additional information about PAF is available on our website: www.rand.org/paf/

---

[1] The ESD is part of the new Information Dominance Systems Center. See https://www.aflcmc.af.mil/WELCOME/Organizations/Electronic-Systems/

This report documents work originally shared with the DAF on September 26, 2024. The draft report, dated September 30, 2024, was reviewed by formal peer reviewers and DAF subject-matter experts (SMEs).

The views expressed in this report are those of the authors and do not reflect the official policy or position of the U.S. Department of Defense (DoD) or the U.S. Government. Review of this material does not imply DoD endorsement of factual accuracy or opinion.

Acknowledgments

We extend our gratitude to Victoria Coleman, Chief Scientist of the Air Force, for sponsoring this project. We also thank Cynthia Lu Schurr, Special Assistant to the Chief Scientist of the Air Force, for her support and guidance throughout the duration of this effort.

This project greatly benefited from the insights provided by representatives from the organizations we investigated for our case studies: Air Force Materiel Command's AFLCMC/HB; Deputy Chief of Staff, Air Force Futures, Headquarters, United States Air Force (HAF A5/7); and Headquarters, Pacific Air Forces (HQ PACAF). Their thoughtful contributions were invaluable in guiding our efforts. We also appreciate the inputs from other stakeholders in offices under the Secretary of the Air Force, Acquisition (SAF/AQ), at Air Force Materiel Command and the Air Force Personnel Center (AFPC).

We are grateful to our RAND colleagues who supported this project: Nelson Lim, Allison Hottes, and Richard Mason. Special thanks to Anna Walkowiak for her project management. Finally, we thank our reviewers, Kirsten Keller and Ginger Groeber, for their constructive comments on the report.

Summary

Issue

Bolstering technical capabilities throughout the Department of the Air Force (DAF)—military and civilian, active and reserve—is a current goal of the department. Although the DAF has many tools to address science, technology, engineering, and mathematics (STEM) needs within its military workforce, there are many barriers to achieving similar goals within the civilian workforce. This project examines the DAF's needs for STEM talent in the civilian workforce, identifies potential gaps in technical competencies, describes options for closing technical skill gaps, and proposes strategies to better track the supply of and demand for STEM talent.

Approach

Because of the size and complexity of the DAF civilian workforce, our analysis centered on three case studies selected in consultation with the project sponsor: the Air Force Life Cycle Management Center, Digital Directorate (AFLCMC/HB; referred to as PEO Digital); the Deputy Chief of Staff, Air Force Futures; and Headquarters, Pacific Air Forces (HQ PACAF). The project team reviewed manpower and personnel data, conducted interviews with subject-matter experts (SMEs) in the case study organizations, analyzed relevant job announcement information, developed and tested two surveys, and conducted a literature scan of training and development opportunities.

Key Findings

What we know about the demand for technical talent includes the following:

- Existing DAF data systems and personnel practices (e.g., job analyses, position descriptions [PDs], competency frameworks) are incomplete and may provide misleading signals of demand for technical talent.
- DAF organizations have distinct mission sets, organizational histories, and sizes, all of which shape civilian billet structures and STEM talent needs.
- The demand signal for technical competencies is localized and not well aligned with strategic workforce functions.
- Case study SMEs identified several competency demands, some spanning organizations and others unique to a particular organization or mission.

What we know about the supply of technical talent includes the following:

- Knowledge of the supply of civilian personnel with specific skills is localized with few methods for tracking them.
- The level of expertise of a particular skill set is often self-reported or not reported at all.

What we know about the gaps in technical talent includes the following:

- Determining gaps in technical talent is complex, and procedures for doing so remain unclear.
- Gaps arise from various causes, including personnel being moved from one program to another, unfunded positions, delays and barriers due to the civilian hiring system, and short-term needs requiring specific expertise.

How the DAF currently address gaps in technical talent includes the following:

- Gaps in STEM talent are addressed using multiple talent pools—contractors, other organizations, and federally funded research and development centers (FFRDCs).
- Organizations focus on hiring recent college graduates with the right soft skills and then invest in their training and development.
- When available, the DAF addresses gaps in technical talent through diverse functional development opportunities.

Cross-cutting issues affecting workforce management include the following:

- Information about the civilian workforce—the demands for and the supply of technical talent—is largely decentralized to local, individual organizations and work units.
- The term *STEM* has multiple definitions and may not be the right unit of analysis for estimating the current and future workforce.
- The civilian personnel system creates barriers to hiring talent, which are not limited to STEM occupations.
- Large language models (LLMs; e.g., Generative Pre-trained Transformer [GPT] 4.0) and machine learning (ML) are useful tools for augmenting human analysis of text documents (e.g., PDs) but are not yet sophisticated enough to replace SMEs.

Recommendations

Focus resources and support at the local level.

- Enhance support for local units through strategic engagement.
- Communicate to supervisors the purpose for tracking technical skills.
- Support local organizations and work units to identify critical technical skills.

Identify and evaluate mechanisms to close specific technical skill gaps.

- Encourage local levels to identify and define their most critical technical competencies.

Consider investing in the building blocks of a talent management system.

- Identify which technical skills need development and assess whether the necessary training is available.
- Enhance data collection about the civilian workforce to generate more complete and accessible data.
- Consider developing a broad DAF-wide competency framework for technical skills.
- Conduct periodic workforce surveys to identify supply, demand, and technical skills gaps.
- Adopt a holistic approach to workforce planning that includes the civilian workforce, uniformed service members, contractors, and personnel from FFRDCs.

Contents

About This Report iii

Summary v

Figures and Tables x

CHAPTER 1 1
Introduction 1
    Study Objective and Approach 2
    Structure of this Report 4

CHAPTER 2 5
Demand for Civilians with Technical Competencies 5
    Potential Sources of Demand Signals 6
    Analysis of National Labor Exchange Job Announcement Data 8
    Conclusion 17

CHAPTER 3 18
Determining the Supply of STEM Civilian Personnel 18
    STEM Civilian Personnel in the Department of the Air Force 18
    Challenges Determining the Supply of Civilian Personnel with Technical Competencies 23
    Conclusion 25

CHAPTER 4 26
Exploring Civilian STEM Needs Through Three Case Studies 26
    Background on the Case Study Organizations 26
    Key Insights from the Case Studies 29
    Conclusion 41

CHAPTER 5 42
Using Surveys for Supply and Demand Assessments of Technical Skills 42
    Designing a Survey: Supply and Demand for Technical Competencies 42
    A Second Survey: Valuing Characteristics for Current and Future Needs 45
    Conclusion 48

CHAPTER 6 49
Strategies for Filling Gaps in STEM Competencies 49
    Enhancing Strategic Use of STEM Pipelines 49
    Competency Focus: Digital Modeling and Engineering 50
    Competency Focus: Artificial Intelligence 55
    Competency Focus: Data Science 61
    Competency Focus: Radar 62
    Conclusion 63

CHAPTER 7 64
Key Findings and Recommendations 64
    Key Findings 64
    Recommendations 68
    Conclusion 71

APPENDIX A 72
Large Language Model Skill and Competency Extraction 72

APPENDIX B 75
Case Study Interview Methodology and Protocol 75

APPENDIX C 77
Supplemental Information on Case Studies and Case Study Populations 77

APPENDIX D 84
Pilot Surveys for Supply and Demand Assessments of Technical Skills 84

APPENDIX E 91
Development of STEM Pipelines and Capabilities 91

APPENDIX F 99
Broad Civilian Workforce Challenges in the Department of the Air Force 99

Abbreviations 103
References 105

Figures and Tables

Figures

Figure 2.1. Occurrences of Artificial Intelligence and Machine Learning Keywords in Industry Job Announcements over Time 11

Figure 2.2. Keyword Counts in Industry Job Announcements over Time: Selected Defense-Related Technologies 14

Figure 2.3. Competency Labels for Artificial Intelligence/Machine Learning in Industry Job Announcements over Time 16

Figure 2.4. Competency Labels for Defense-Related Technologies in Industry Job Announcements over Time 17

Figure 3.1. STEM and Non-STEM Civilian Employees 18

Figure 3.2. STEM Civilian Employees by STEM Occupational Series 19

Figure 3.3. STEM and Non-STEM Department of the Air Force Civilians by Grade (End of FY 2023) 20

Figure 3.4. STEM Civilian Employees by Pay Plan (End of FY 2023) 21

Figure 5.1. Air Force Futures Survey Results, Averages 46

Figure 5.2. Air Force Futures Survey Results, Current and Future Changes 47

Figure C.1. Percentage of PEO Digital Personnel by Occupational Group 78

Figure C.2. Percentage of PEO Digital Personnel by Grade 78

Figure C.3. Percentage of PEO Digital Personnel by Highest Degree 79

Figure C.4. Percentage of Headquarters, Pacific Air Forces Personnel by Occupational Group 81

Figure C.5. Percentage of Headquarters, Pacific Air Forces Personnel by Grade 81

Figure C.6. Percentage of Headquarters, Pacific Air Forces Personnel by Highest Degree 82

Figure C.7. Percentage of Headquarters, Pacific Air Forces Personnel by Academic Specialty of Highest Degree 83

Tables

Table 2.1. Potential Demand Signals in the Department of the Air Force 6

Table 2.2. Number of Job Announcements by Engineering Specialization 10

Table 2.3. Occurrences of Artificial Intelligence and Machine Learning Keywords in Department of the Air Force and Industry Job Announcements 10

Table 2.4. Occurrences of Data Analytics Keywords in Department of the Air Force and Industry Job Announcements 12

Table 2.5. Differences in the Frequency of Keywords Used by the Department of the Air Force and Industry ... 12

Table 2.6. Frequency of Defense-Related Technology Keywords Used by the Department of the Air Force and Industry 13

Table 2.7. Twenty Most Frequent Competency Labels Generated by ChatGPT, per Group 15

Table 3.1. Degree Levels for STEM Occupational Groups (End of FY 2023) 21

Table 3.2. Top Ten Academic Degree Specialties and STEM Versus Non-STEM Specialties for Engineering and Architecture Occupational Group 22

Table 4.1. Interviews Conducted with the Three Case Study Organizations 26

Table 6.1. Systems Engineering Research Center Digital Engineering Competency Framework 51

Table 6.2. Examples of Digital Engineering–Focused Programs or Courses 53

Table 6.3. Department of Defense Artificial Intelligence Education Strategy Competency List 56

Table 6.4. Department of Defense Artificial Intelligence Competency Curriculum Topics at Three Levels of Depth 59

Table A.1. Prompts Used in Large Language Models–Augmented Job Announcement Analysis 73

Table C.1. Air Force Futures Civilian Positions 79

Table D.1. PEO Digital Survey Questions—Part 1: Requirements 85

Table D.2. PEO Digital Survey Questions—Part 2: Hiring Challenges 86

Table D.3. Relevant Survey Definitions 87

Table D.4. Proficiency Scale 88

Table D.5. Risk Scale: Level of Proficiency 88

Table D.6. Risk Scale: Number of Personnel 89

Table D.7. Air Force Futures Survey: Valuing Characteristics in Civilian Hires 89

Table E.1. Student STEM Pipeline Programs 92

Table E.2. Core Certification Institutions 96

Table F.1. Types of Personnel Covered by Selected Programs 101

Chapter 1

Introduction

The U.S. Department of Defense (DoD) is in the midst of a comprehensive effort to posture the force for future conflicts, as evidenced by the 2022 National Defense Strategy (NDS) focus to "act urgently to sustain and strengthen deterrence, with the People's Republic of China as our most consequential strategic competitor and the pacing challenge for the Department."[1] In line with NDS defense priority four, "building a resilient Joint Force and defense ecosystem,"[2] DoD published its *National Defense Science & Technology Strategy 2023* with the assertion that "we need to recruit, retain, and engage the most talented people in the world—both those in our workforce today and in the workforce of the future."[3] More recently, the Department of the Air Force (DAF) announced plans to streamline and bolster technical capabilities.[4]

Central to this effort is fielding a force equipped with education, training, and skills in science, technology, engineering, and mathematics (STEM). To maintain technological advantage, engage complex problem sets, and support warfighter needs, the DAF is prioritizing STEM to drive new innovations across all domains and in the joint environment. As part of this effort, the Secretary of the Air Force tasked the Air Force Chief Scientist (AF/ST) to assess the posture and capacity of the DAF to provide organic technical excellence needed for strategic competition. This tasking evolved into the Secretary of the Air Force's Management Initiative 9, which generated several recommendations to increase investments in STEM capabilities for military personnel.[5]

The DAF has a range of tools and levers to address STEM needs within its military workforce, from increasing billet requirements for advanced STEM degrees to purposefully managing military STEM career fields.[6] Indeed, every officer who commissions from the U.S. Air Force Academy receives a Bachelor of Science degree, and the Academy can adjust coursework internally to increase institutional focus on specific technological training capabilities for its future officer workforce. Yet, in terms of DAF civilian management, efforts to yield a robust STEM workforce face challenges ranging

---

[1] DoD, "Fact Sheet: 2022 National Defense Strategy," undated.

[2] DoD, undated.

[3] DoD, *National Defense Science & Technology Strategy 2023*, May 9, 2023.

[4] Secretary of the Air Force Public Affairs, "Air Force, Space Force Announce Sweeping Changes to Maintain Superiority Amid Great Power Competition," February 12, 2024.

[5] Special Assistant to the Air Force Chief Scientist, "Management Initiative MI #9 Update," May 4, 2022.

[6] Aleah M. Castrejon, "AFRL Team Works to Boost Number of Advanced STEM Degrees," August 15, 2022.

from structural constraints embedded within federal recruiting and hiring practices for civilians,[7] a small civilian talent pool from which to recruit desired skill sets,[8] difficulties retaining civilians due to lack of advancement opportunities,[9] coupled with private-sector competition,[10] and a hierarchical military structure that discourages the innovation and autonomy sought by many civilian STEM professionals.[11]

With the military services experiencing ongoing recruitment challenges,[12] DAF investments in the civilian STEM workforce offer opportunities to maximize technical expertise and personnel continuity within units and across organizations. Though constrained by burdensome federal regulations, DAF civilian hiring efforts also enable direct accessions of skilled technical talent into all grade levels. Additionally, civilians who previously worked in the private sector may help the DAF increase partnerships with key industry partners, thus contributing to a resilient defense ecosystem as outlined in the aforementioned NDS defense priority four.[13]

Study Objective and Approach

In fiscal year (FY) 2024, AF/ST asked RAND Project AIR FORCE (PAF) to examine the DAF's civilian workforce needs for STEM technical talent, identify potential gaps in workforce size and competencies, and explore strategies to strengthen the STEM workforce. Given the significant contributions that the civilian workforce[14] can make to enhancing the DAF's technical capabilities, this project evaluated the demand and supply signals for increased civilian STEM talent across the DAF.

Considering the size and complexity of the DAF civilian workforce, our initial step was to identify specific DAF areas for focused analysis. In collaboration with the project sponsor, we selected the

---

[7] Ginger Groeber, Paul W. Mayberry, Brandon Crosby, Mark Doboga, Samantha E. DiNicola, Caitlin Lee, and Ellen E. Tunstall, *Federal Civilian Workforce Hiring, Recruitment, and Related Compensation Practices for the Twenty-First Century: Review of Federal HR Demonstration Projects and Alternative Personnel Systems to Identify Best Practices and Lessons Learned,* RAND Corporation, RR-3168-OSD, 2020.

[8] Kirsten M. Keller, Ginger Groeber, Philip Armour, Jenna W. Kramer, Shirley M. Ross, Diana Y. Myers, Hannah Acheson-Field, Samantha E. DiNicola, Shreyas Bharadwaj, and Stephanie Williamson, *Attracting and Employing Top-Tier Civilian Technical Talent in the Department of the Air Force: A Comparison of Six Occupations with Other Federal Agencies and the Private Sector,* RAND Corporation, RR-A987-1, 2023.

[9] Kirsten M. Keller, Maria C. Lytell, David Schulker, Kimberly Curry Hall, Louis T. Mariano, John S. Crown, Miriam Matthews, Brandon Crosby, Lisa Saum-Manning, Douglas Yeung, Leslie Adrienne Payne, Felix Knutson, and Leann Caudill, *Advancement and Retention Barriers in the U.S. Air Force Civilian White Collar Workforce: Implications for Demographic Diversity,* RAND Corporation, RR-2643-AF, 2020.

[10] Ginger Groeber, Kirsten M. Keller, Philip Armour, Samantha E. DiNicola, Irina A. Chindea, Brandon Crosby, Ellen E. Tunstall, and Shreyas Bharadwaj, *Department of the Air Force Civilian Compensation and Benefits: How Five Mission Critical and Hard-to-Fill Occupations Compare to the Private Sector and Key Federal Agencies,* RAND Corporation, RR-A334-1, 2021.

[11] Shirley M. Ross, Rebecca Herman, Irina A. Chindea, Samantha E. DiNicola, and Amy Grace Donohue, *Optimizing the Contributions of Air Force Civilian STEM Workforce,* RAND Corporation, RR-4234-AF, 2020.

[12] David Barno and Nora Bensahel, "Addressing the U.S. Military Recruiting Crisis," *War on the Rocks,* March 10, 2023.

[13] DoD, undated.

[14] For a review of RAND research on the civilian workforce in the DAF and other military services, see "RAND Research on Civilian Workforce Issues," webpage, May 14, 2024.

following three distinct organizations, each differing in mission, size, structure, and type of STEM requirements:

- Air Force Materiel Command's Air Force Life Cycle Management Center, Digital Directorate (AFLCMC/HB), referred to as Program Executive Office Digital and Enterprise Services (PEO Digital, now the Electronic Systems Directorate, or ESD[15]);
- Deputy Chief of Staff, Air Force Futures, Headquarters, U.S. Air Force (HAF A5/7)
- Headquarters, Pacific Air Forces (HQ PACAF).

In consultation with the project sponsor, we deliberately excluded organizations traditionally viewed as STEM-centric, such as the Air Force Research Laboratory (AFRL), because of its already extensive focus on STEM and technical talent. Our objective was to examine organizations closer to the operational environment, which AF/ST identified as still requiring significant technical expertise. Additionally, we sought to include a variety of organizational levels in the case studies in order to provide a comprehensive overview of the differing STEM needs across various segments of the DAF.

Although these organizations are comprised of personnel from a wide range of occupational series, discussions and analyses focused on the following three occupational groups defined by the U.S. Office of Personnel Management (OPM):

- Engineering and Architecture Group (0800 series), which includes general engineering and all engineering disciplines
- Mathematical Sciences Group (1500 series), which includes operations research, data science, mathematics, and statistics
- Physical Sciences Group (1300 series), which includes physical sciences, chemistry, meteorology, and geological disciplines.

The project team conducted a comprehensive analysis of the demand for civilian technical talent within the case study organizations and occupational groups. This involved examining the education and skill requirements for key roles, as well as assessing the number of personnel needed to fulfill these requirements.[16] The methods used for this task included reviewing manpower data, conducting interviews with subject-matter experts (SMEs), and analyzing relevant job announcement information.

Next, the team established a baseline understanding of the existing supply of technical personnel across the case study organizations. This workforce analysis combined an examination of personnel data, such as occupational series and education levels, with targeted discussions and interviews with SMEs.

The research then turned to identifying the key challenges and gaps faced by the case study organizations in building the technical workforce they require. Through in-depth discussions with SMEs, the team gathered feedback on critical technical competencies as well as broader issues within the federal civilian personnel system.

---

[15] The ESD is part of the new Information Dominance Systems Center. See https://www.aflcmc.af.mil/WELCOME/Organizations/Electronic-Systems/

[16] We use the term *skill* broadly throughout this report to include technical knowledge, skills, and abilities (KSAs). We use the term *competency* to refer to one or more sets of KSAs.

Recognizing the limitations of existing systems in providing sufficient demand and supply information, the project team developed and tested two survey types. The first type was designed for work units that directly support the development or maintenance of technologies. The second type was designed for work units that may have a less defined need for technical talent but may benefit from employees having an awareness or basic foundational knowledge of different technologies. These surveys included questions to uncover needed technical competencies, the demand for these competencies, and the current supply of civilians with these technical competencies. The team conducted pilot tests of these surveys and discussed the results with SMEs to refine the approach. Another objective of the survey development was to equip the DAF with a systematic method for collecting information on the supply of, demand for, and potential gaps in STEM skills.

Finally, the team conducted a thorough literature scan to explore possible mechanisms for addressing any deficiencies that were uncovered. This review investigated a range of options, including university programs, industry partnerships, and continuing education opportunities, with the aim of informing potential solutions.

Using all the information gathered, the project team developed recommendations for the DAF to prioritize efforts that help to position the DAF to more clearly identify current and emerging needs and to evaluate current workforce capabilities.

Structure of this Report

The results of these various analyses are presented in the following chapters. Chapters 2 and 3 examine the demand for STEM competencies in the civilian workforce and the supply of civilian personnel with STEM talent, respectively. After introducing the case study organizations, Chapter 4 delves into the insights gained from discussions with representatives from these organizations, with an emphasis on their needs for civilian personnel with STEM competencies and how they fulfill those needs. Following from the case studies, Chapter 5 explores the use of surveys as a tool for gathering information on the supply and demand of technical skills. Chapter 6 focuses on how to fill gaps in STEM talent by examining four specific competencies: digital modeling and engineering, artificial intelligence (AI), data science, and radar. The report concludes in Chapter 7 with a summary of findings and the project team's recommendations.

The report contains six supplemental appendixes. Appendix A describes the methodology for using large language models (LLMs) in the analysis of demand for civilian personnel with STEM skills. Appendix B provides the interview methodology and protocol used for our case study evaluation, and Appendix C provides additional details on the case study organizations and populations. Appendix D contains the surveys completed as part of the case studies. Appendix E presents background information on the DAF's development and implementation of STEM pipelines and capabilities, and Appendix F assesses broad DAF civilian workforce challenges.

Chapter 2

Demand for Civilians with Technical Competencies

The call for more STEM talent is a recurring theme throughout DoD and the DAF. In 2010, a National Research Council report examining U.S. Air Force (USAF) STEM workforce needs assessed that "force reductions, ongoing military operations, and budget pressures are creating new challenges for attracting and managing personnel with the needed technical skills."[1] The report recommended that the USAF recruit, develop, employ, and retain STEM skills and experiences as key enabling objectives to maintaining and advancing core competencies in technology and cyber operations.[2]

A 2014 RAND report identified significant unmet STEM needs across USAF career fields, as well as finding that little attention was paid to reviewing future STEM academic degree requirements in relation to supporting air, space, and cyberspace operations.[3] Additionally, the establishment of the U.S. Space Force (USSF) in 2019 raised new demands to field a workforce skilled in STEM disciplines.[4]

Despite calls for more STEM, there is little practical insight into what "more STEM" means, which makes it challenging for the DAF to identify the best actions to expand its technical workforce. Previous RAND research comparing compensation and benefits for STEM workers in the federal government and in the private sector found inconsistencies across existing research in how the STEM workforce is defined.[5] Moreover, there are multiple dimensions comprising the STEM workforce, from degree and education level to occupation, geography, and career stage.[6]

At present, the DAF has not defined what technical talent means, which is a barrier for hiring civilians into such positions. This problem extends beyond the DAF to DoD, where the lack of

---

[1] National Research Council of the National Academies, *Examination of the U.S. Air Force's Science, Technology, Engineering, and Mathematics (STEM) Workforce Needs in the Future and Its Strategy to Meet Those Needs*, 2010.

[2] National Research Council of the National Academies, 2010.

[3] Lisa M Harrington, Lindsay Daugherty, Craig Moore, and Tara L. Terry, *Air Force-Wide Needs for Science, Technology, Engineering, and Mathematics (STEM) Academic Degrees*, RAND Corporation, RR-659-AF, 2014.

[4] Michael Spirtas, Yool Kim, Frank Camm, Shirley M. Ross, Debra Knopman, Forrest E. Morgan, Sebastian Joon Bae, M. Scott Bond, John S. Crown, and Elaine Simmons, *Creating a Separate Space Force: Challenges and Opportunities for an Effective, Efficient, Independent Space Service*, RAND Corporation, RB-10103-AF, 2020.

[5] Kathryn A. Edwards, Maria McCollester, Brian Phillips, Hannah Acheson-Field, Isabel Leamon, Noah Johnson, and Maria C. Lytell, *Compensation and Benefits for Science, Technology, Engineering, and Mathematics (STEM) Workers: A Comparison of the Federal Government and the Private Sector*, RAND Corporation, RR-4267-OSD, 2021.

[6] Edwards et al., 2021.

specificity regarding what DoD means by "technical talent" introduces challenges to filling gaps in such talent, if they are present.[7] Solving this challenge requires the DAF to answer several key questions:

- What is the definition of STEM, from a workforce perspective?
- What are the specific technical competencies that the DAF requires?
- What methodology should the DAF use to determine its technical competencies?

In this chapter, we review possible data sources that may provide information on the demand for specific technical competencies and explore various methods for tracking emerging skill demands, with a series of analyses to extract competency information from job announcements. By tracking competencies across specific DAF organizations, industry, and over time, we may gain insights into which competencies are currently in demand and which may become more important in the future.

Potential Sources of Demand Signals

Multiple sources of information can offer insights into the demand for STEM skills within the DAF. However, the information in these sources is frequently incomplete, stored in decentralized locations, lacks specificity, or inaccurate. Table 2.1 provides an overview of potential demand signals for STEM skills within the DAF. It includes a brief description of various sources of information that might indicate the need for specific competencies and skills, such as occupational series, manpower authorizations, job analyses, competency management frameworks, performance appraisals, job announcements, and training requests.

The table also describes several limitations associated with the various sources of information used to gauge STEM skill demands within the DAF. These limitations include a lack of specificity, as many sources do not indicate position-specific competencies or KSAs, which in turn makes assessing precise requirements difficult. In addition, some job categories lack their own occupational series, and unfunded manpower authorizations may hide true demand, leading to incompleteness in the data. Decentralized and inconsistent usage of competency management frameworks and systems further complicates the issue, as these frameworks are not fully implemented or consistently used across different work units, and there is no standardized terminology.

Table 2.1. Potential Demand Signals in the Department of the Air Force

Source	Description	Primary Limitation(s)
Occupational series	This is a classification system grouping positions with similar duties, responsibilities, and qualifications into specific categories (e.g., Operations Researcher series).	• Many different work roles can be found within a single occupational series. • The series does not indicate position-specific competencies or KSAs (e.g., digital modeling). • Some job categories lack their own occupational series (e.g., systems engineering).

---

[7] Diana Gehlhaus, Maria C. Lytell, James Ryseff, and Kirsten M. Keller, *Keeping Up with the Joneses: How Can DoD Address Its Technical Talent Shortage?* RAND Corporation, PT-A2884-3, 2023.

Source	Description	Primary Limitation(s)
Manpower authorizations (funded and unfunded)	Authorizations are specifications of the human resources required for a work unit's mission. Funded authorizations are approved positions with allocated budgets for hiring and paying employees, while unfunded authorizations lack current budget allocation.	• Authorizations may not accurately reflect the specific skills or grades required for actual job duties. • Unfunded authorizations may skew or hide the true demand for particular skills. • Static authorizations may not adapt quickly to changing mission requirements and thereby fail to capture emerging skill needs. • Manpower authorizations may not align with how the work is accomplished (e.g., authorizations accounted for by office symbol versus a work breakout structure system).
Job analyses	A job's duties, responsibilities, and qualifications are systematically analyzed to determine necessary skills and attributes for effective performance. The job analysis worksheet aids the process by organizing and documenting this information.	• It is unclear if and how often analyses are conducted. • Information gathered focuses on tasks performed, with less emphasis on KSA requirements.
Competency management frameworks and systems	These document the skills and numbers needed and assess the current workforce's skill sets. They are developed for different purposes and by different organizations to include Air Force Materiel Command and acquisition functional communities.	• Frameworks and system in place are not fully implemented or consistently used (i.e., some work units do not utilize them). • There is no standardized terminology for describing KSA requirements across career fields or within a career field working across different organizations (e.g., major commands [MAJCOMs]).
Performance appraisals	These evaluate an employee's job performance and productivity.	• Information may not accurately reflect actual job duties. • Evaluations focus more on outcomes than on job activities or KSAs.
Job announcements	These official notifications provide information about job openings and corresponding job title, duties, responsibilities, and other required qualifications.	• A single job announcement can have multiple positions. • Competencies and KSAs listed are often not specific to the position. • Position announcements may not be easily accessible or centralized across the different personnel systems (e.g., AcqDemo, Lab Demo, Defense Civilian Intelligence Personnel System [DCIPS]).
Position descriptions (PDs)	An official document describes an employee's job duties and corresponding KSAs required to perform those job duties.	• Duties listed on the PD may not reflect the actual work being performed. • Details about KSAs may not be specific to a position.
Requests for training	These record requests for centrally or unit-funded functional education or training. This training is prioritized as critical or mandated, essential or recommended.	• Training requests may be influenced by available funding, which can lead to uncommunicated needs when funds are unavailable or allocated elsewhere. • Systems that capture requests do not include no-cost or low-cost sources such as Air University and Defense Acquisition University (DAU).

NOTE: AcqDemo = Department of Defense Civilian Acquisition Workforce Personnel Demonstration Project; Lab Demo = DoD Science and Technology Laboratory Demonstration Project.

Static and outdated information is another concern, with manpower authorizations not adapting quickly to changing mission requirements and job analyses not always being conducted regularly. Misalignment with actual duties is also problematic, since performance appraisals and manpower authorizations may not accurately reflect the specific skills required. In addition, job and position announcements may not be easily accessible or centralized across different personnel systems. Finally, if funding is unavailable or allocated elsewhere, the need for STEM skills or training may not be communicated. Collectively, these limitations illustrate why it is so difficult to obtain a comprehensive and accurate understanding of the demand for STEM skills within the DAF.

To better understand the implications of these limitations as they relate to discerning STEM needs from job announcements specifically, we explored the potential of leveraging an external source of job announcements in conjunction with AI and machine learning (ML) techniques. The idea is that skill demands for an organization should be reflected, in part, by the requirements listed in job announcements used to attract applicants to specific roles. Identifying and tracking skill requirements over time could be a viable approach to determining emerging demands for the DAF and to facilitating comparisons with industry standards. In the following sections, we discuss the promise and limitations of using job announcements to evaluate and compare competencies across the DAF and private industry. Although this method shows potential, it remains an imperfect solution, and making such comparisons is complicated.

Analysis of National Labor Exchange Job Announcement Data

The National Labor Exchange (NLx) is a partnership between the National Association of State Workforce Agencies and the Direct Employers Association. It serves as a comprehensive and publicly accessible online job search engine that aggregates job postings from various sources, including employer websites, state job banks, and federal government job portals. By providing a centralized platform for job seekers and employers and ensuring that job postings are current, accurate, and widely disseminated, NLx aims to enhance the efficiency of the labor market.

NLx maintains an extensive database of position announcements that encompass a wide range of industries and job categories. The data include detailed information about job titles, descriptions, required qualifications, skills, and competencies. The database is continuously updated to reflect the latest job openings, thereby offering a dynamic and up-to-date repository of labor market information. The data can be leveraged for various analytical purposes, such as tracking trends in job requirements, comparing skill demands across different sectors, and identifying emerging competencies in the workforce.

NLx was chosen for this analysis due to its comprehensive coverage, public accessibility, and transparency of sources for job posting extraction. However, potential limitations include sampling bias and gaps in coverage, as NLx may not capture all job postings, especially from nonparticipating employers. Furthermore, the data may not cover certain positions in the DAF such as those with Direct Hire Authority if they were not posted on USAJOBS. Despite these issues, NLx provides a robust dataset suitable for analyzing trends. The choice of NLx was also influenced by cost considerations and accessibility, as private datasets can be prohibitively expensive.

In this section we present a text-based analysis of position announcements drawn from the NLx database. The data compiled were from September 2015 to January 2024 and included 119,364,668 announcements across a broad variety of occupations. We then used keywords and metadata filters to narrow the data to 15,809 announcements for defense-related engineering positions. The goal of our analyses is to find evidence of demand signals in job announcements for DAF and industry engineering positions using keywords and competency descriptions. We explored two types of demand signals: the frequency of specific keywords and competencies over the past ten years and the emergence or change in frequency of keywords and competencies over time. The purpose of this analysis is not to set an absolute value or threshold to indicate when a demand is present, but to offer insights on the relative frequency over time and between the DAF and industry. Our first analysis involves a keyword search for terms and phrases associated with emerging technologies or core competencies of defense-related engineering positions. Our second analysis uses a large language model to augment the extraction and labeling of competencies from job announcements.

Keyword Search

To search for demand signals for specific technical competencies in NLx data using keyword searching, we employed the following methodology:

- **Data collection and grouping:** We used a collection of Occupational Information Network (O*NET) occupation series codes representing various engineering subfields,[8] such as electrical engineering and mechanical engineering, along with defense-related keywords and NLx metadata variables. This allowed us to create two comparable sets of job announcements— one for the DAF and one for industry. The first set comprised all job announcements for civilian engineering positions within the DAF captured by NLx from the end of FY 2015 through the first part of FY 2024. We observed that within this time frame, DAF engineering positions were predominantly concentrated between FY 2016 and FY 2020, with few entries from FY 2021 to FY 2024. The industry set consisted of randomly sampled job announcements for defense-related engineering positions at companies that are federal contractors, with compilation dates evenly distributed across FY 2016 to FY 2024. Although the date ranges do not directly align, we included more recent data for industry job announcements to capture emerging technology skill sets such as AI and ML. Each group contained 2,520 entries. Table 2.2 shows the distribution of specializations by O*NET occupation series code in the two groups. The definitions for role, such as electrical engineer are based on O*NET classifications within the NLx data.
- **Keyword compilation:** We compiled a collection of keywords representing skills, competencies, and knowledge areas from case study interviews and the Critical and Emerging Technologies List Update Report.[9] Several groups of these keywords were associated with specific emerging technologies, such as *hypersonics* or *artificial intelligence*, while others were more general, such as those related to technology awareness.

---

[8] The O*NET website contains "hundreds of standardized and occupation-specific descriptors on almost 1,000 occupations covering the entire U.S. economy. The database, which is available to the public at no cost, is continually updated from input by a broad range of workers in each occupation." See https://www.onetcenter.org/overview.html

[9] Fast Track Action Subcommittee on Critical and Emerging Technologies, *Critical and Emerging Technologies List Update*, National Science and Technology Council, 2022.

- **Keyword analysis:** For each keyword, we counted the number of job announcements containing the keyword in their descriptions. We controlled for case sensitivity and punctuation. Each job announcement was counted as unique in the analysis if it had unique job identifiers according to the NLx data variables, even if the job announcements were similar or identical.

Table 2.2. Number of Job Announcements by Engineering Specialization

	DAF Group	Industry Group
Mechanical Engineer	528	423
Aerospace Engineer	606	296
Electronics Engineer	567	263
Industrial Engineer	None	763
Electrical Engineer	178	481
Civil Engineer	399	None

We summarize the counts and trends in Tables 2.3 through 2.6 using four groups of keywords: AI and ML, data analytics, keywords illustrating different emphases in job announcements, and defense-related technologies. The results indicated that the frequency of keywords varied across DAF and industry position announcements. For example, there were only ten occurrences of the AI keyword for the DAF but 127 occurrences for industry (Table 2.3). Assuming that position announcements are a relevant source of demand, this finding suggests there is some demand for AI competencies among engineering specializations in industry but no corresponding demand in the DAF. It is important to note that the absence of keywords from job announcements does not necessarily indicate a lack of demand, but could reflect variations in terminology or differences between what is needed at the time of hiring versus what may be performed after gaining further experience and training.

Table 2.3. Occurrences of Artificial Intelligence and Machine Learning Keywords in Department of the Air Force and Industry Job Announcements

Keyword	DAF	Industry
"artificial intelligence"	6	56
"AI"	3	4
"machine learning"	1	41
"ML"	0	12
"AI/ML"	0	11
"deep learning"	0	3
"reinforcement learning"	0	0

The occurrence of AI and ML keywords suggest that some industry positions are seeking skill sets related to these emerging technologies. Figure 2.1 illustrates the emergence of these same keywords in industry job announcements over time, expressed as a percentage of the total number of job announcements in the sample for each fiscal year. The sharp increase of relevant keywords in the job announcements, demonstrated in Figure 2.1, indicates a more enduring demand, which we refer to as a *demand signal*—in this case for AI and ML.

Figure 2.1. Occurrences of Artificial Intelligence and Machine Learning Keywords in Industry Job Announcements over Time

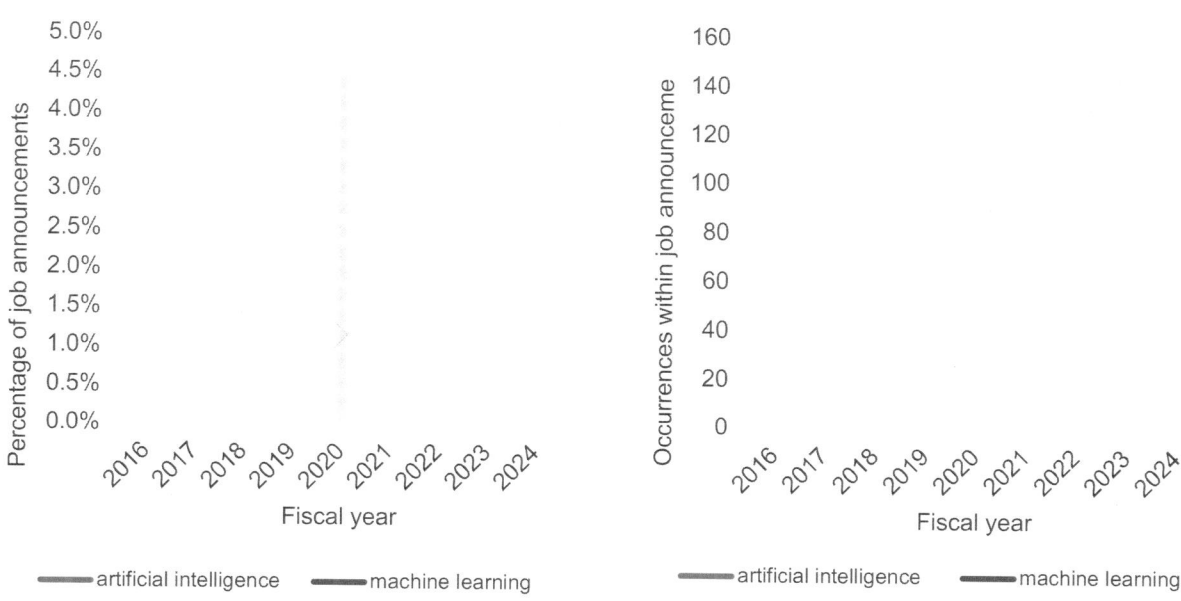

NOTE: Red is machine learning, blue is artificial intelligence. The green dashed line marks FY 2020, the fiscal year when Generative Pre-trained Transformer 3 (GPT-3) was released.

As noted previously, different organizations may use varying terminology to describe competency requirements. For example, Table 2.4 shows that in our sample of NLx data, the phrase "data processing" appears more frequently in DAF job announcements than it does in industry. Conversely, industry job announcements more commonly use the keyword "data analysis" than DAF job announcements.

It is unclear whether "data processing" in DAF job announcements refers to skills comparable to "data analysis" or other data-related skills in industry. DAF job announcements may be selecting for comparable data analytic skills through other means, such as education requirements or more general terminology. For example, the phrase "computer science" appeared in 940 DAF job descriptions, compared with 249 industry job announcements. Additionally, DAF engineering roles may emphasize different skill sets compared with their industry counterparts, even if both are for engineering positions. Table 2.5 shows the frequency of keywords that highlight some observed differences in terminology between DAF and industry job announcements.

Table 2.4. Occurrences of Data Analytics Keywords in Department of the Air Force and Industry Job Announcements

Keyword	DAF	Industry
"data analysis"	79	156
"data analytics"	2	23
"data visualization"	0	7
"data evaluation"	34	1
"data processing"	444	28
"data storage"	1	2

Table 2.5. Differences in the Frequency of Keywords Used by the Department of the Air Force and Industry

Keyword	DAF	Industry
"program management"	**516**	197
"programming"	**753**	242
"systems engineering"	**553**	358
"emerging tech"	**367**	60
"materials"	**2,003**	579
"automation"	8	**226**
"characterization"	17	**111**
"communications"	213	**581**
"thermal"	7	**202**
"firmware"	8	**68**
"5G" or "6G"	0	**26**

NOTE. Numbers are bolded to highlight the column with greater frequency.

DAF job announcements may emphasize different skill sets than industry does. If DAF's strategy is to hire talent with less experience but with the ability to learn through on-the-job training, job announcements may advertise for general skill sets rather than more specific competencies that can be developed over time. Additionally, the DAF often uses the same job announcement to cover multiple positions, which may necessitate advertising for more general requirements to attract candidates for various roles. DAF job announcements might also contain different keywords because the requirements for DAF civilian engineering positions differ from those in industry. Considering these potential differences between the DAF and industry, we caution that the frequency and specificity of keywords may not accurately represent true skill demands.

Lastly, we searched for keywords related to defense technologies, such as "hypersonics." Table 2.6 shows the frequency of these keywords across sample groups, with a selection mapped over time

Table 2.6. Frequency of Defense-Related Technology Keywords Used by the Department of the Air Force and Industry

Keyword	DAF	Industry
"hypersonic"	8	28
"hypersonic propulsion"	1	5
"space propulsion"	2	7
"propulsion systems"	34	36
"radar"	136	144
"RF" or "radio frequency"	114	302
"spectrum management"	28	1
"lasers"	4	24
"control"	1,519	1,191
"detection"	44	115
"tracking"	164	158
"distribution"	122	204
"payload"	14	97
"sensor"	102	277
"transmission"	354	94
"instrument"	210	269
"navigation"	88	85
"timing (PNT)"	1	4
"UAS," "UAV," or "USV"	59	32

NOTE: PNT = positioning, navigation, and timing; UAS = unmanned aerial system; UAV = unmanned aerial vehicles; USV = unmanned surface vehicle.

in Figure 2.2. Generally, more specific keywords, such as "hypersonic propulsion" or "spectrum management," appear less frequently less specific keywords such as "control" or "radar."

While the frequency of certain keywords may signal some kind of change in demand, additional analyses are needed to determine the specific skill sets required. For example, Figure 2.2 shows a sharp increase in the keyword "detection" in our industry sample from FY 2023 to FY 2024, compared with "rf," "lasers," and "hypersonic." However, without further context, it is unclear whether "detection" refers to a specific technology or multiple technologies with overlapping needs for detection.

In our sample, the increase in "detection" from FY 2023 to FY 2024 reflected a confluence of demands for detection skill sets in various applications such as radiation detection, intrusion and alarm systems, and signals analysis. Even in this case, where the demand signal can be parsed by domain, it remains unclear whether the relative increase of detection-related job announcements is due to overlapping terminology or an increase in demand for similar detection skill sets across domains.

Figure 2.2. Keyword Counts in Industry Job Announcements over Time: Selected Defense-Related Technologies

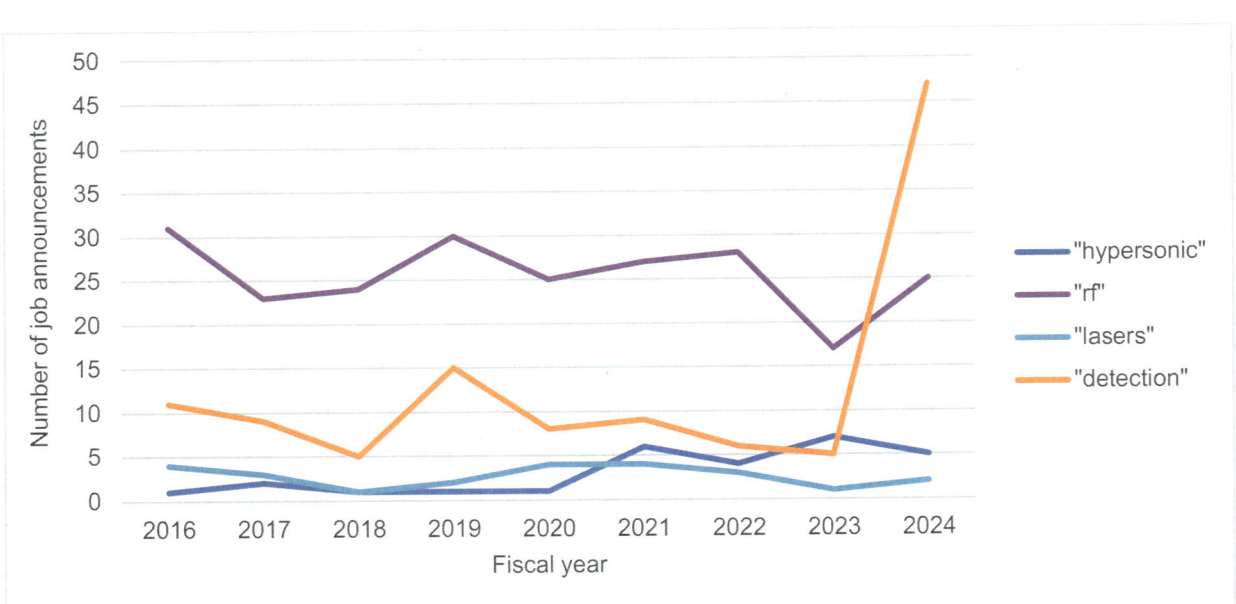

Large Language Models–Augmented Skill and Competency Extraction

In the previous section, we described performing a keyword analysis to gain insights into the demand for emerging technologies such as AI or defense-related technologies such as hypersonics and spectrum management. While the frequency of AI and ML keywords may signal an emerging need for AI and ML competencies, there was less certainty in which skill sets or competencies are needed when exploring other sets of keywords. Therefore, we used an LLM, OpenAI's Generative Pre-Trained Transformer 4.0 (GPT-4), on the same data and groups to explore methods for addressing this limitation.

Using an LLM, we extracted keywords and phrases from job announcements, determined whether they represented technical (e.g., STEM skills) or nontechnical competencies (e.g., soft skills), and then used the LLM to produce labels describing the technical competencies. Our hypothesis was that the LLM would produce labels that better describe the required skill sets or competencies, thereby improving the quality of demand signals observed through keyword counts. Since GPT-4 does not have a standardized competency bank, we expected that labels would vary across similar job requirements. To help address this, we adopted a workflow that grouped similar labels and job requirements, then standardized them using the most common competency label for each group. (See Appendix A for a detailed description of our methodology.) To summarize the results of this analysis, Table 2.7 lists the top 20 labels identified for the DAF and private industry.

In Figure 2.3, we explore demand signals for AI and for defense-related technologies using the competency labels generated by GPT for the industry group. Trends from competency labeling are consistent with the trends we saw in keyword analysis. Each term represents a group of competency labels—for example, the AI/ML group includes "AI and Machine Learning Concepts," "AI Technologies," "Machine Learning," "Predictive Forecasting," and "Autonomous Robotics and AI Development." Similar to the keyword analysis trends shown in Figure 2.1 for AI/ML, the competency labeling method also captures an increasing trend in AI/ML competency labels.

Table 2.7. Twenty Most Frequent Competency Labels Generated by ChatGPT, per Group

Group	Term	Count
DAF	Construction and Engineering	475
	Database Management	428
	Electronics Engineering	398
	Systems Engineering	**396**
	Engineering Principles and Practices	321
	Test and Evaluation	287
	Multidisciplinary Engineering	285
	Engineering Concepts and Practices	250
	Weapon Systems	245
	Engineering Knowledge and Practices	227
	Systems Integration	216
	Civil Engineering	206
	Maintenance and Engineering	206
	Design and Technology Assessment	197
	Site/Facility Survey and Inspection	197
	Mechanical Engineering	**192**
	Industrial Engineering	182
	Aerospace Engineering	**175**
	Quantitative Analysis	154
	Design Criteria Evaluation	142
Industry	Electronics Engineering	436
	Industrial Engineering	354
	Mechanical Engineering	**319**
	Manufacturing	213
	Systems Engineering	**195**
	Test Method Development	182
	Requirements Engineering	167
	Technical Documentation	158
	Data Analysis	157
	Machine Troubleshooting	156
	3D CAD Design	152
	Materials Science	148
	Hardware and Software Design	145
	Mechanical Design	145
	Statistical Quality Control	144
	Statistical Analysis	135
	Root Cause Analysis	132
	Aerospace Engineering	**131**
	Test Equipment Operation	131
	Equipment Maintenance	127

NOTE: Bolded rows indicate terms that appeared in the top 20 for both DAF and industry.

Figure 2.3. Competency Labels for Artificial Intelligence/Machine Learning in Industry Job Announcements over Time

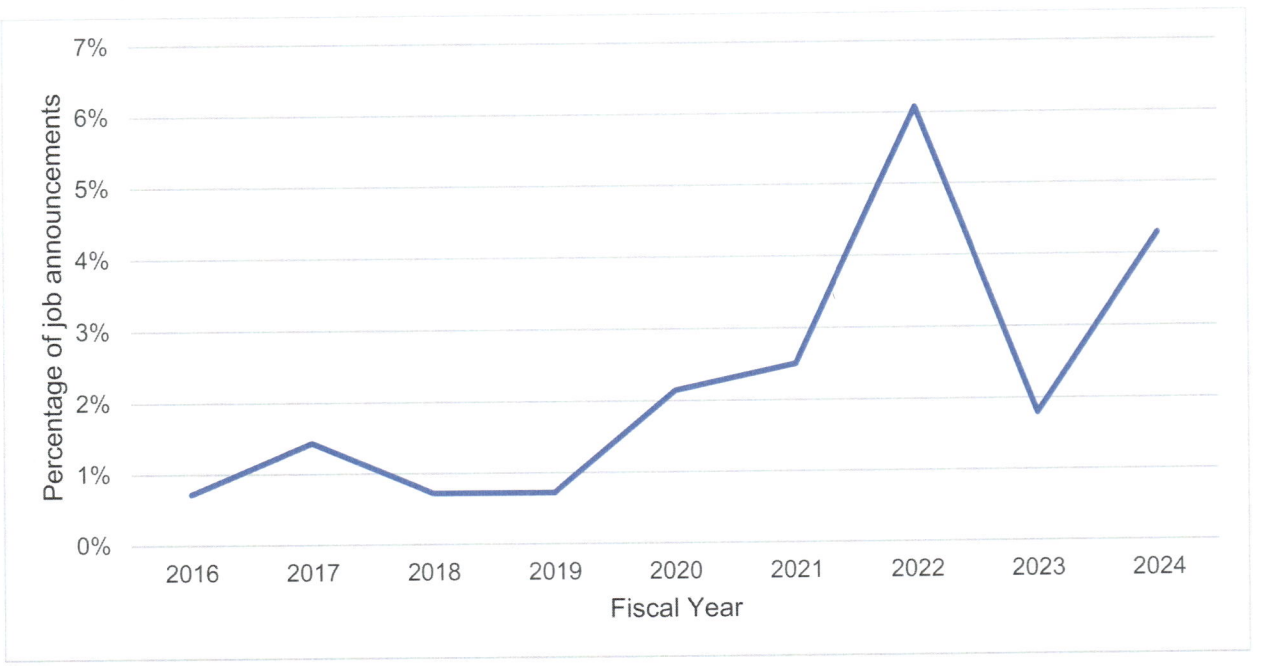

Like the trendlines in Figure 2.2, the pattern of the trendlines for defense-related technologies in Figure 2.4 represented by hypersonic, RF, laser, and detection are also similar. However, the LLM method does not show as sharp an increase in number of detection-based labels from FY 2023 to FY 2024 as it does in the keyword search for "detection" in the same time range. One possible explanation for this difference is that the LLM methodology uses the full PD tasks to determine different meanings for "detection"; whereas the keyword analysis searches only for the occurrence of a word or set of words.

In addition to keyword trend analysis, the LLM methodology allowed us to calculate the density of technical competencies in job announcements. For each job announcement, we calculated the ratio of the number of competency labels extracted by GPT-4 to the total length of the job announcement, calculated from the data in the "description" variable of the NLx dataset.

On average, the ratio of technical competency labels to job announcement length for DAF positions was .002, compared with .016 for industry job announcements. There was an average of six technical competency labels extracted from a DAF job announcement, and the average length of an average DAF job announcement was 3,058 words. For industry job announcements, the averages were 11 labels and 798 words, respectively. The competency labels extracted for industry job announcements were also more specific than their DAF counterparts: After standardization within groups, GPT-4 extracted 1,546 unique competency labels from industry job announcements and 461 unique competency labels from DAF job announcements. This pattern suggests that DAF announcements were longer but were less specific and contained fewer skill requirements.

16

Figure 2.4. Competency Labels for Defense-Related Technologies in Industry Job Announcements over Time

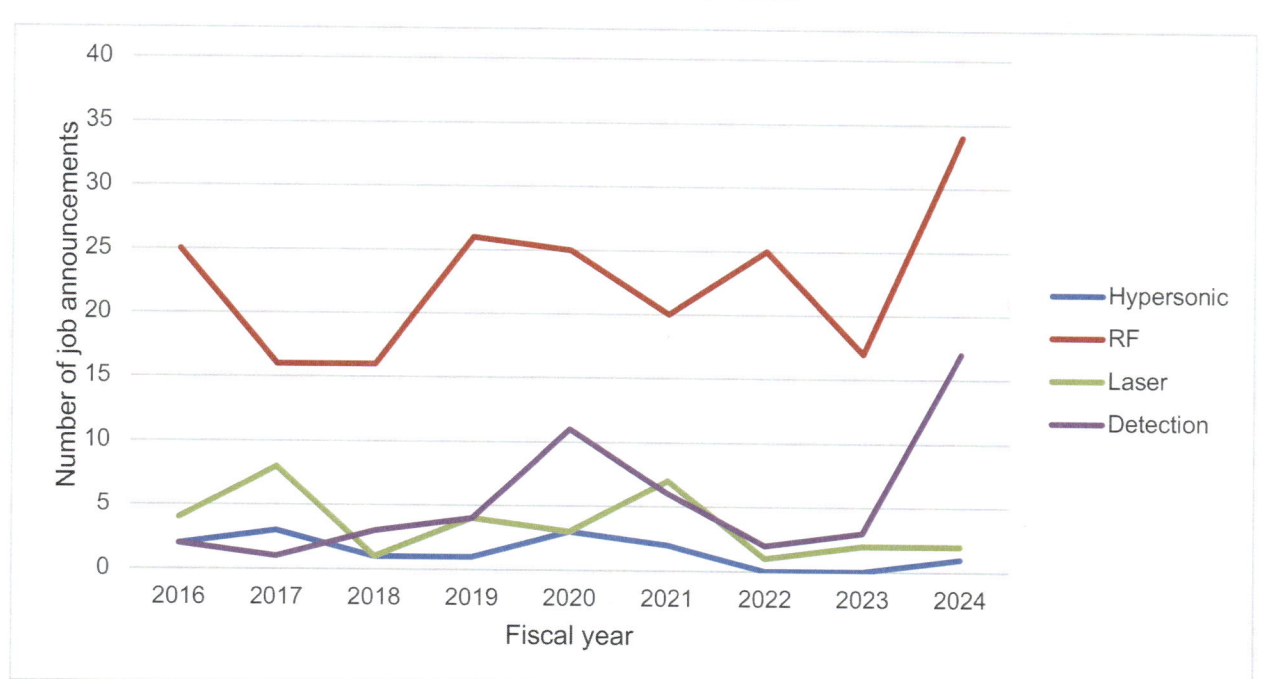

While our LLM-based analysis complements the keyword trends and demand signals from job announcements, our methodology is not a perfect characterization of technical skill demands. One reason for this is that GPT's mapping of job announcements to competency labels was not entirely accurate, which we describe in further detail in the validation section of Appendix A. There was also a considerable amount of judgment required to standardize competency labels, which may have removed demand signals for specific competencies in favor of signaling more general technical competencies.

Conclusion

Advances in LLMs and ML methods may help improve the accuracy and comprehensiveness of competency labeling to allow for better resolution of demand signals in job announcement data. But ultimately, the quality of demand signals in job announcement data will be limited by the representation of desired skill sets and competencies included in job announcements. Nonetheless, investing in a workflow that combines natural language processing (NLP), LLMs, and SME inputs can be an effective first step toward summarizing the prevalence of technical competencies contained in job announcements or other potential data sources (see Table 2.1). The usefulness of this type of workflow will increase as gaps in data about DAF position duties and KSA requirements are closed.

Determining the Supply of STEM Civilian Personnel

In this chapter, we present a survey of the overall supply of STEM civilian personnel in the DAF and include comparisons with non-STEM personnel. Following a presentation of that data, we discuss the challenges in understanding the supply of STEM personnel in terms of the presence of specific competencies in any population.

STEM Civilian Personnel in the Department of the Air Force

Figure 3.1 illustrates the number of personnel in STEM occupational series (0800, 1500, and 1300) compared with those in non-STEM occupational series from FY 2012 through FY 2023. During this period, the percentage of STEM civilian personnel in the DAF increased from 13.4 percent to 15.9 percent. Personnel in these STEM occupational series constitute a relatively small portion of the overall DAF civilian workforce. However, this does not imply that personnel

Figure 3.1. STEM and Non-STEM Civilian Employees

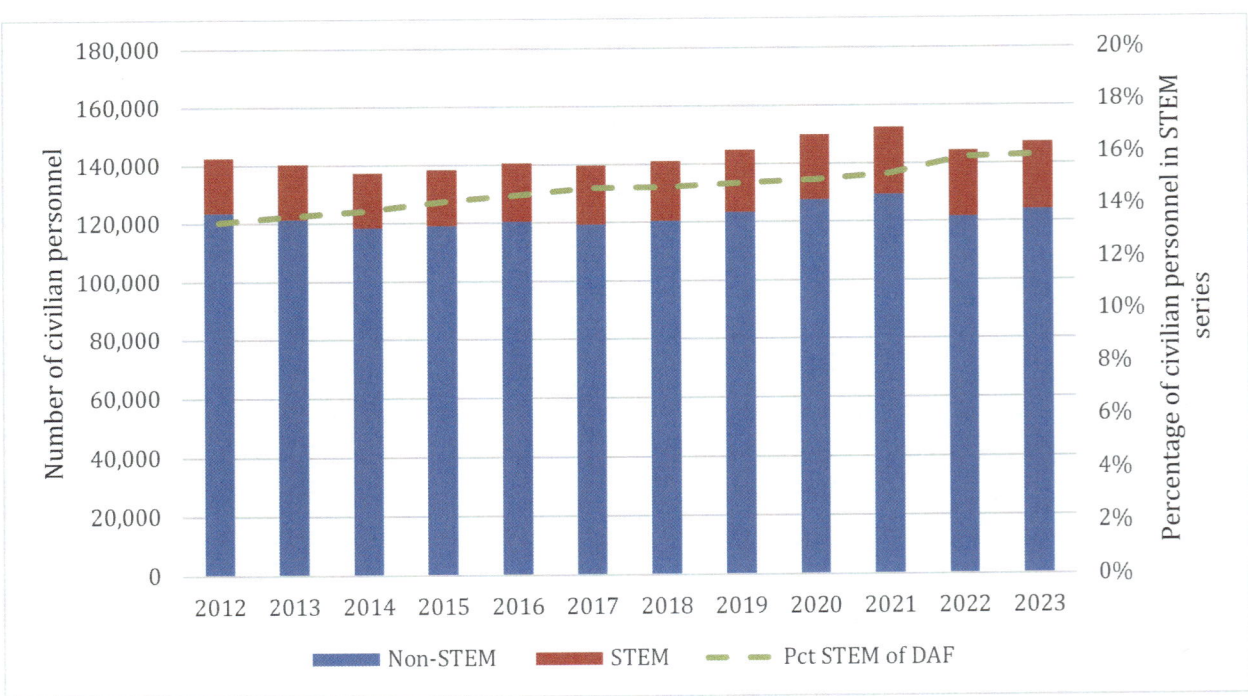

SOURCE: RAND analysis of Defense Civilian Personnel Data System (DCPDS), as of September 2023.

in non-STEM occupational series lack technical competencies. Individuals in non-STEM series may possess technical skills from previously holding STEM positions, earning degrees in STEM disciplines, or gaining experience through specific technical duties while serving in non-STEM roles.

Figure 3.2 presents the number and percentage of STEM civilian personnel in each STEM occupational series by fiscal year. The number of civilian personnel in the engineering occupational series (0800) is consistently greater than in the mathematical (1500) and physical sciences (1300) series. However, the percentage of personnel in the engineering series has declined steadily for the past decade, falling from 83 percent in FY 2012 to 76 percent in FY 2023. Conversely, the mathematical sciences occupational series has grown from 1,893 personnel in FY 2012 to 4,010 personnel in FY 2023. Our discussions with organizational leaders indicated that this increase reflects a heightened emphasis on data sciences and data analysis.

Figure 3.2. STEM Civilian Employees by STEM Occupational Series

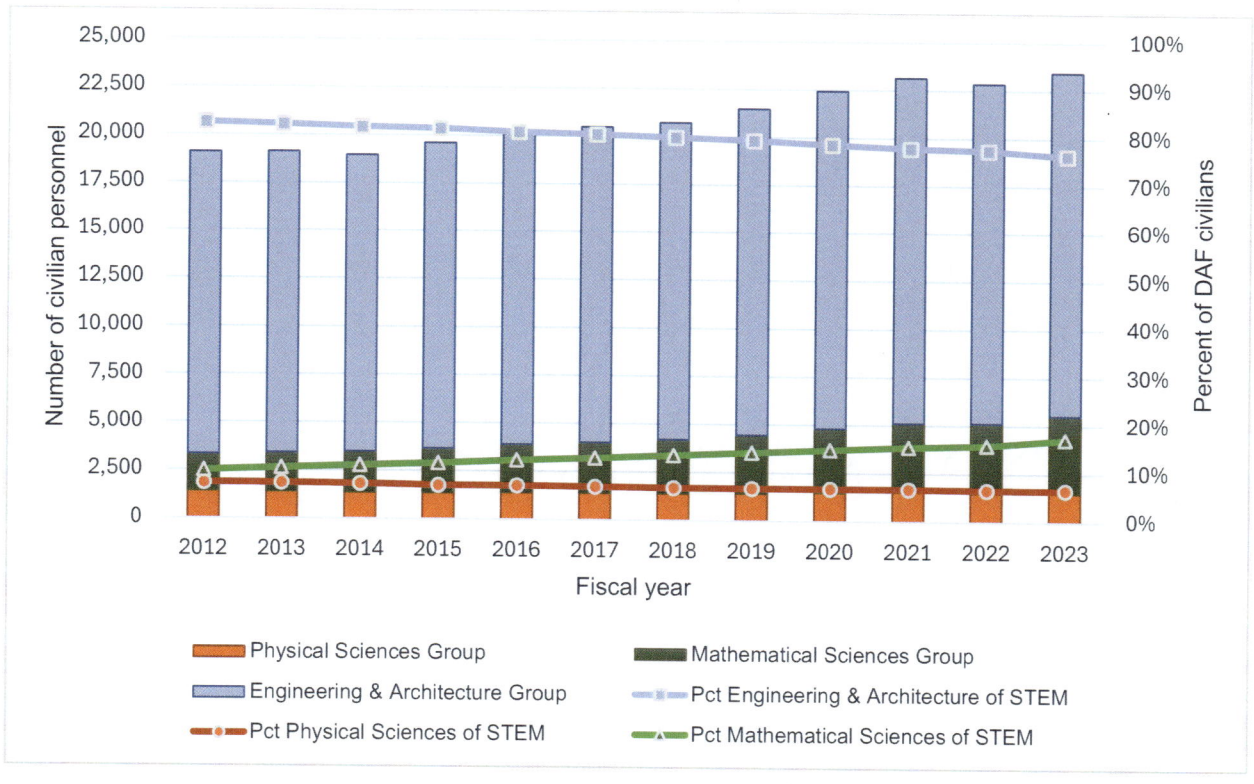

SOURCE: RAND analysis of DCPDS, as of September 2023.
NOTE: STEM occupational groups are physical sciences (1300), mathematical sciences (1500), and engineering (0800).

Figure 3.3 compares civilian employees by grade in the engineering, physical sciences, and mathematical sciences occupational groups with all other occupational groups. Generally, employees in these STEM occupational series tend to be more senior compared with those in the rest of the DAF. This higher-graded, and consequently more expensive, workforce has implications for setting compensation to attract and retain sufficient civilians in these technical series.

Figure 3.3. STEM and Non-STEM Department of the Air Force Civilians by Grade (End of FY 2023)

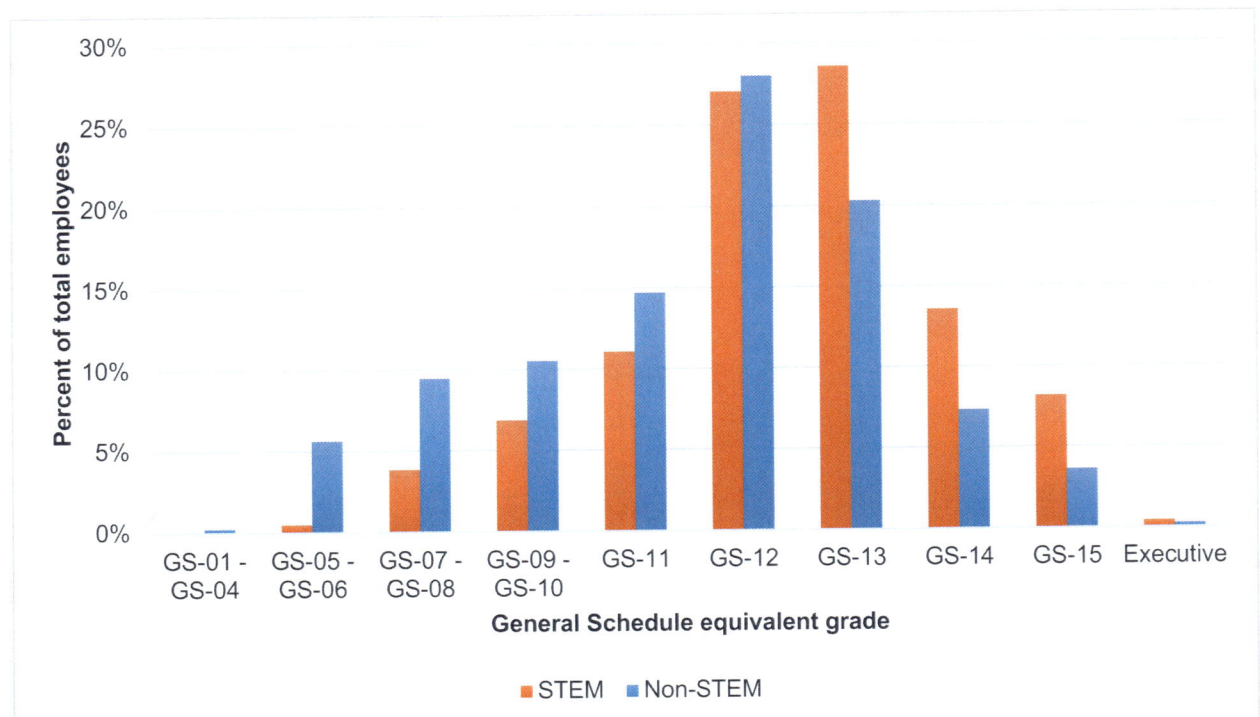

SOURCE: RAND analysis of DCPDS, as of September 2023.
NOTE: STEM is defined as the engineering (0800), physical sciences (1300), and mathematical sciences occupational groups. Grades across all pay plans are categorized according to the equivalent General Schedule (GS) grade.

Another dimension of the supply of STEM personnel in the DAF is the pay plan under which they are employed. Managing STEM personnel in the DAF is complicated by various hiring and pay-setting authorities. Different organizations may use these tools differently, giving some units an advantage in attracting and retaining high-demand STEM workers, which can impede a senior manager's ability to allocate talent where needed. The traditional federal personnel system, governed by Title 5 of the U.S. Code, offers limited flexibility with employees hired under the General Schedule (GS) classification and pay system. This system mandates "equal pay for substantially equal work" and requires a competitive hiring process, often seen as rigid.

Personnel systems such as AcqDemo, Lab Demo, and the DCIPS offer more flexibility with pay bands and contribution-based compensation. While these tools benefit certain organizations, they can create disparities among co-located organizations, giving some an advantage in attracting and retaining STEM workers. This poses challenges for managers in recruiting and retaining talent effectively in that those using traditional systems may face constraints compared with those using more flexible tools. Appendix F discusses the challenges in managing personnel under differing pay and personnel systems.

Figure 3.4 shows employees in STEM occupational groups by their GS-equivalent grade and the specific pay plan under which they fall. Seventy percent of STEM civilians in GS-equivalent grades of GS-12 through GS-15 are under AcqDemo or Lab Demo, yet there are also significant numbers of STEM civilians in these grades under the traditional GS pay system.

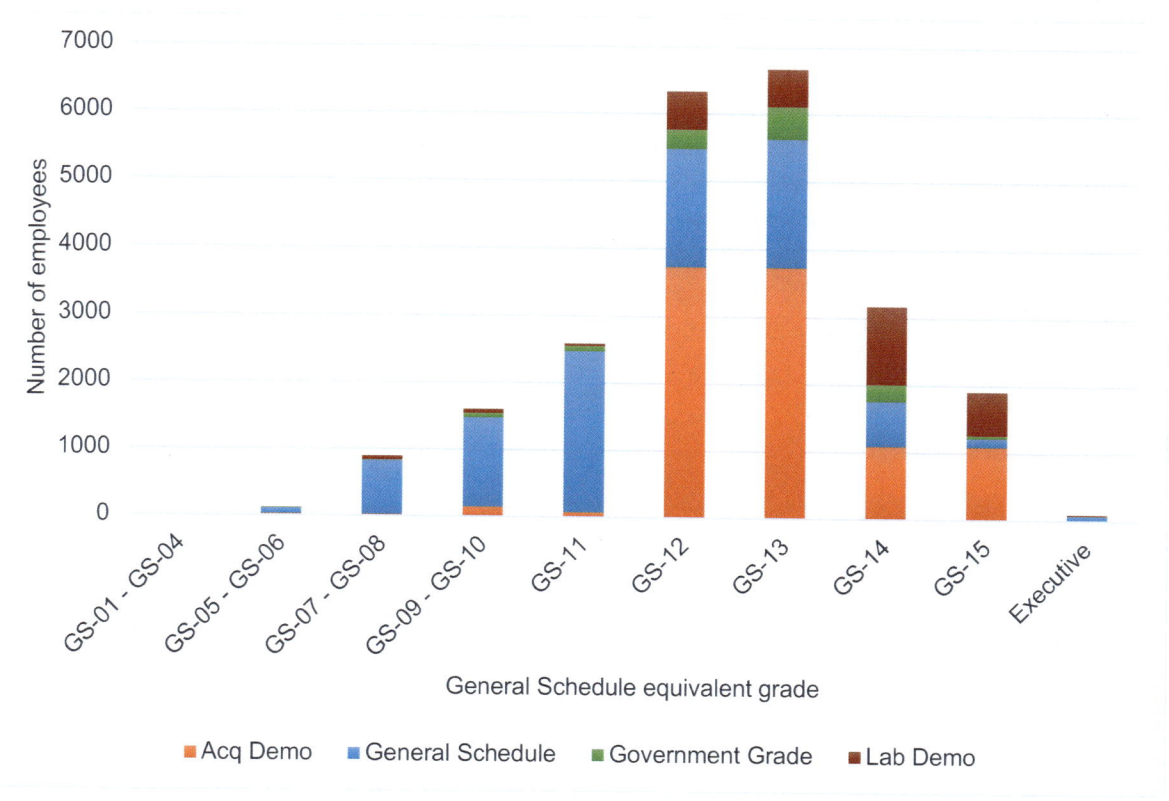

Figure 3.4. STEM Civilian Employees by Pay Plan (End of FY 2023)

SOURCE: RAND analysis of DCPDS, as of September 2023.
NOTE: STEM is defined as the engineering (0800), physical sciences (1300), and mathematical sciences occupational groups. Grades across all pay plans are categorized according to the equivalent GS grade. Personnel who fall under the Government Grade pay plan are managed primarily under DCIPS.

Another way to analyze the supply of STEM civilian personnel is by their educational levels and academic specializations. Table 3.1 presents the highest degree earned by civilians in each of the STEM occupational groups.

Table 3.1. Degree Levels for STEM Occupational Groups (End of FY 2023)

	Occupational Group			
	Engineering and Architecture	Physical Sciences	Mathematical Sciences	Total
Total personnel	17,886	1,536	4,010	23,432
Degree level (percentage of total in group)				
Ph.D.	5%	24%	5%	6%
Master's or professional	38%	30%	39%	38%
Bachelor's	42%	34%	55%	44%
Below bachelor's	15%	13%	1%	12%

NOTE: STEM occupational groups are physical sciences (1300), mathematical sciences (1500), and engineering (0800).

Table 3.2 shows the top ten academic specialties for the largest STEM occupational group, Engineering and Architecture. This table highlights some challenges associated with using academic levels and specialties recorded in personnel records as indicators of STEM skills within the civilian workforce. First, 9 percent of personnel in this group have no specialty listed for their highest degree; hence, there is limited information on the skills they may have acquired through their education. This gap can be partially attributed to the fact that academic degrees and specialties are often self-reported or only documented to the extent necessary for job qualification.

Table 3.2. Top Ten Academic Degree Specialties and STEM Versus Non-STEM Specialties for Engineering and Architecture Occupational Group

		Number of Civilian Personnel	Percentage of Total Civilians in 0800
Top ten academic specialties (highest degree earned)	Electrical, Electronics, and Communications Engineering	3,735	21
	Mechanical Engineering	2,692	15
	None listed	1,620	9
	Aerospace, Aeronautical, and Astronautical/Space Engineering	1,304	7
	Business Administration, Management and Operations	1,169	7
	Computer Engineering	713	4
	Civil Engineering	604	3
	Engineering, General	443	2
	Systems Engineering	353	2
	Engineering-Related Fields	348	2
Total in 0800 occupational group	STEM degrees	13,618	76
	Non-STEM degrees	2,638	15
	No degree specialty listed	1,630	9
	Total civilian personnel	17,886	

NOTE: Data as of the end of FY 2023. Degree specialties are categorized according to the Department of Education's Classification of Instructional Programs (National Center for Education Statistics, "The Classification of Instructional Programs," webpage, undated). Degrees are classified as STEM or non-STEM according to the Department of Homeland Security's STEM Designated Degree Program List (U.S. Department of Homeland Security, "DHS STEM Designated Degree Program List," webpage, undated).

Second, 15 percent of personnel hold a non-STEM degree as their highest degree, with Business Administration, Management, and Operations appearing in the top ten. These personnel may have completed other degree programs or have technical qualifications that indicate STEM skills. However, the fact that these additional qualifications may not be easily accessible or in personnel data complicates the assessment of STEM competencies within the workforce. Several stakeholders we interviewed shared this view; one branch chief specifically noted that to accurately determine the degrees his personnel hold, he needed to ask them directly.

Academic degrees may be insufficient indicators of the technical skills individuals possess for reasons other than the limitations of personnel records described above. First, degrees may emphasize theoretical knowledge, which may not translate directly into practical skills needed in the workplace. Hands-on experience, internships, and real-world projects may be needed to develop practical expertise. One interviewee related that there may be people with just the skills they need, but not a degree. "They joke about the kid hacking from their parents' basement with the skills they have but don't have the degree. Maybe traditional degrees don't work but the OPM requirements are constraints."

Second, the skills taught in academic programs may not always align with current industry demands or emerging technologies. Rapid advancements in technology can render some academic knowledge outdated by the time graduates enter the workforce. Additionally, academic programs may provide a broad overview of a field but might not delve deeply into specialized skills required for specific roles. Conversely, they might focus on a narrow specialization and miss out on broader competencies.

Third, the quality and rigor of academic programs can vary significantly among institutions. Consequently, the same degree from different schools may not represent the same level of competency. The dynamic field of technology demands continuous learning, and an academic degree represents a snapshot of knowledge at a specific point in time. Certain technical skills are best validated through certifications and specialized training programs that are more focused and up to date than traditional academic degrees.

Challenges Determining the Supply of Civilian Personnel with Technical Competencies

This overview of the supply of technical civilian personnel within the DAF offers useful insights, but the information is incomplete and lacks finer details. Obtaining more detailed information and assessing the supply of personnel with STEM competencies in the DAF civilian workforce presents challenges. Based on interviews and a review of previous research on civilian management and STEM technical talent, we identify several additional challenges in determining both the current and projected supply of civilian personnel with technical competencies.[1]

Data availability and quality pose substantial obstacles in accurately assessing STEM competencies. A critical limitation lies in the federal personnel system, which typically classifies skills at the broad occupational series level. This system lacks the granularity to capture specific technical competencies within these series, creating a significant information gap. This is especially true for cyber skills.[2] For instance, an occupational series might encompass a wide range of skills, but not all individuals within that series will possess the same technical competencies, such as an engineer with digital engineering (DE) skills or a computer programmer with AI/ML skills. In addition, personnel may have acquired specific technical skills in previous roles that are not reflected in their current classification, which further complicates the assessment.

---

[1] Appendix F provides information on broad civilian workforce challenges and prior civilian personnel STEM research.

[2] Martin C. Libicki, David Senty, and Julia Pollak, *Hackers Wanted: An Examination of the Cybersecurity Labor Market*, RAND Corporation, RR-430, 2014, p. 62.

To overcome these limitations, researchers and managers often rely on individual self-assessments, which can introduce potential biases and inaccuracies. Personnel must be directly queried about their specific skills, and the organization must depend on the accuracy of these self-reported competencies. As one manager explained, "This method is fraught with challenges, as self-assessments can vary widely in reliability." Moreover, personnel records often suffer from incompleteness, outdated information, or inaccuracies, which further hamper efforts to evaluate the current state of technical skills within the workforce. The lack of standardization in data collection and categorization across different organizations and work units leads to additional inconsistencies, compounding the challenges in analysis and comparison efforts.

The **definition and scope of STEM competencies** themselves present another layer of complexity. STEM fields encompass a broad spectrum of disciplines, and the specific competencies required can vary significantly across roles and over time. The evolving nature of STEM fields means that what constitutes a STEM job is not static and requires continuous reassessment. Technological advancements and changing industry needs can make it difficult for organizations to keep their competency frameworks up to date. In addition, the increasingly interdisciplinary nature of modern roles, which often combine STEM with other competencies, further complicates the classification process and makes it challenging to maintain clear categorizations.

Rapidly changing technology adds to the difficulty of supply assessment. Rapid technological advancements can swiftly alter the skills and competencies required, thus making it challenging to maintain an up-to-date inventory of available talent.[3] For example, the emergence of new technologies such as quantum computing or advanced materials science can create gaps in existing workforce assessments, since these novel areas may not be adequately captured by current classification systems. The pace of change necessitates continuous monitoring and updating of competency frameworks to ensure organizations can respond to technological changes effectively.

Retention and attrition factors can have a significant impact on the stability of the supply of technical talent. High turnover rates—for example in a particular high-demand specialty such as cybersecurity—can lead to fluctuations in the available pool of talent, making it difficult to maintain an accurate picture of the workforce's capabilities. The loss of experienced personnel not only reduces the number of available experts but also means there are fewer skilled and experienced personnel to mentor and contribute to the development of newer employees, especially when work units depend on on-the-job training for new employees. All this can potentially create a knowledge gap.

Recruitment and training also affect the supply of STEM-competent personnel. The inability to attract and develop new talent with the necessary STEM skills can affect the overall supply and potentially create shortfalls in needed critical skills. The DAF faces competition from the private sector, which often offers more competitive salaries and benefits, thus making it increasingly difficult to attract and retain top STEM talent.[4] In addition, the length and complexity of the federal hiring process can deter potential candidates. Training and development programs within the DAF must be

---

[3] Libicki, Senty, and Pollak, 2014.

[4] Edwards et al., 2021.

robust and adaptive to ensure that personnel can acquire and maintain relevant competencies, but these programs require significant investment and coordination.[5]

Internal mobility within the DAF adds another layer of complexity to understanding the supply of STEM personnel with particular skills. Personnel movements between roles and promotions can shift the distribution of STEM competencies across and within organizations. For instance, an individual with advanced AI/ML skills may move into a managerial role in which those technical skills are less directly applied. Furthermore, ongoing cross-training and skill development programs can alter the skill sets of existing personnel, rendering static assessments less reliable and necessitating frequent reevaluations. The dynamic nature of career progression and skill acquisition means there is a need for continuous skill evaluations to keep pace with changing job requirements.[6]

Finally, **policy and budgetary constraints** play a significant role in shaping the STEM workforce. Funding limitations can have an impact on the ability to hire and develop STEM personnel, which in turn can potentially create gaps in critical areas. For example, budget cuts can lead to reduced training opportunities or delayed hiring. Furthermore, shifts in policy priorities can affect the resources allocated to STEM workforce development and thus lead to fluctuations in supply and demand dynamics over time. Changes in national security priorities or technological focus areas can redirect attention and funding away from certain STEM fields, affecting the overall competency landscape.

Conclusion

In summary, accurately determining the supply of STEM-competent personnel within the DAF requires a comprehensive and adaptive approach that incorporates continuous data collection and strategic workforce planning to ensure that the DAF can meet its current and future STEM needs.

---

[5] Chapter 6 provides information on strategies for filling gaps in STEM competencies.

[6] LinkedIn, *2018 Workplace Learning Report: The Rise and Responsibility of Talent Development in the New Labor Market*, 2018.

Chapter 4

Exploring Civilian STEM Needs Through Three Case Studies

As previous chapters discuss, information on the supply and demand for STEM civilian talent in the DAF is limited. To address this information gap, we selected different DAF organizations, in coordination with our sponsor's office, as case studies to explore STEM supply and demand in more detail. These organizations—PEO Digital, Air Force Futures (HAF A5/7), and HQ PACAF—may have very different civilian workforces and diverse requirements for technical skills. To capture this complexity across different types of DAF organizations, we conducted interviews with SMEs to understand the nuances shaping supply and demand dynamics for civilian STEM needs. For this analysis, we defined STEM personnel as individuals in three occupational groups: Engineering (0800), Physical Sciences (1300), and Mathematical Sciences (1500).[1] Our analysis of these case studies produced several key insights into how the DAF may consider identifying and filling civilian STEM requirements within and across organizations.

Background on the Case Study Organizations

Using a semi-structured protocol, we administered a total of 27 interviews across the three case study organizations, as shown in Table 4.1.[2] SMEs participating in the interviews were

Table 4.1. Interviews Conducted with the Three Case Study Organizations

Organization	Office Symbol	Perspective	Number of Interviews
PEO Digital		Sustainment and acquisition of technology	15
Airborne Warning and Control System	AFLCMC/HBS		
Aerospace Management Systems	AFLCMC/HBA		
Theater Battle Control	AFLCMC/HBD		
Aerospace Dominance Enabler	AFLCMC/HBZ		
Air Force Futures	HAF A5/7	Headquarters (strategy)	6
Headquarters Pacific Air Forces	HQ PACAF	MAJCOMs	6

---

[1] OPM, *Handbook of Occupational Groups and Families*, December 2018.

[2] See Appendix B for details on the interview protocol and Appendix C for further information on the workforce in the three case study organizations.

supervisors working at different levels of the organization ranging from branch to headquarters and Air staff.[3]

PEO Digital

PEO Digital is a key component of USAF's acquisition and technology ecosystem. Its mission is to "integrate, automate, and accelerate operations to make the world safer for the Joint Force, our allies, and the nation."[4] PEO Digital includes various divisions and teams dedicated to specific projects and initiatives focusing on a wide range of missions aimed at enhancing national defense through advanced command, control, intelligence, surveillance, and reconnaissance capabilities. These enhancements include modernizing and integrating airborne early warning systems, command and control platforms, and air operations centers.

PEO Digital is comprised of a diverse mix of military personnel, civilian employees, and contractors who advance cutting-edge technologies in force protection, strategic deterrence, and mission planning to ensure readiness and superiority in air combat. We focused our efforts on a subset of four divisions and mission areas within PEO (Table 4.1). As of September 2023, there were 168 STEM civilian personnel in our focal PEO Digital divisions, comprising roughly 24 percent of all civilian personnel across those divisions. These STEM personnel tended to be in higher pay grades than non-STEM personnel and had higher educational attainment—with a higher percentage having attained a graduate degree (60.9 percent) than non-STEM personnel (47.2 percent).

Our discussions with PEO Digital provided us with an overview of the directorate's technologies and mission areas, as well as competency gaps and other pain points. In addition, to identify specific workforce demands, including competency requirements and gaps, we gathered inputs from technical experts (i.e., chief engineers) at the branch level.

Air Force Futures

We also evaluated the supply and demand dynamics of civilian STEM talent in Air Force Futures. Previously known as Headquarters Air Force (HAF) Deputy Chief of Staff for Strategy Integration and Requirements, Air Force Futures represents a headquarters-level organization fulfilling a strategic and operational mission set for the DAF. Air Force Futures

> focuses on developing Air Force Strategy and concepts, conducting strategic assessments of the operating environment through wargames and workshops, manifesting an integrated future force design, and achieving timely and effective operational capabilities required for tomorrow's Airmen to fight and win.[5]

According to the Air Force Futures 2022 capability development guidebook, the

---

[3] There are multiple organizational layers within the DAF: (1) service (i.e., the USAF, the USSF), (2) command/headquarters/ air staff (e.g., PACAF), (3) center/wing (e.g., Lifecycle Management Center), (4) directorate/group (e.g., PEO Digital), (5) division/squadron (e.g., AFLCMC/HBA—Aerospace Management Systems), (6) branch/flight (e.g., HBAW—Weather Systems Branch), and (7) section/element.

[4] Air Force Life Cycle Management Center, "Digital Directorate—About Us," undated.

[5] Air Force Biography, "Lieutenant General S. Clinton Hinote," October 2023.

Air Force Futures teams collaborate with strategists and futurists in the Joint Staff, Combatant Commands, Major Commands, Space Force, and intelligence communities to "identify the need" for how the Air Force, as part of the Joint Force, will fight and win in future conflicts.[6]

Air Force Futures comprises three centers: Concepts and Strategy, or Center 1; Capability Development, or Center 2; and Force Design, Integration, and Wargaming, or Center 3. Center 1 aims to describe a family of concepts to capture the future warfighting vision for the Air Force and thus inform force design as well as planning, programming, budgeting, and execution. Center 2 focuses on defining and operationalizing mission needs, and Center 3 works to create an integrated force design for future capabilities to combine and fight together within a future Air Force family of systems. Center 2 uses cross-functional teams and functional integration teams to bridge the gaps between the aspirational concepts from Center 1 and the force design from Center 3. About 14 percent of the Air Force Futures civilians are assigned to traditional STEM career fields.

Our discussions with Air Forces Futures provided insights into civilian personnel requirements, utilization, management, hiring, and STEM needs. We also obtained materials on Air Force Futures organizational structures and mission sets, which we used as supplemental background information to assess and scope our findings from the interviews and from other secondary research for this project.

Headquarters, Pacific Air Forces

Rounding out our case study assessments, we evaluated civilian technical needs within Pacific Air Forces (PACAF), a MAJCOM headquartered at Joint Base Pearl Harbor-Hickam, Hawaii. PACAF's mission is to provide ready air and space power to promote U.S. interests in the Indo-Pacific region. This command is responsible for air operations in an area that spans over 100 million square miles, extending from the west coast of the United States to the east coast of Africa, and from the Arctic to the Antarctic.

PACAF's primary missions include air superiority, global precision attack, rapid global mobility, global integrated intelligence, surveillance and reconnaissance, and command and control. The command plays a crucial role in deterring aggression, assuring allies, and maintaining stability in the Indo-Pacific region. PACAF operates a variety of aircraft, including fighters, bombers, tankers, and reconnaissance platforms, and maintains a robust presence through a network of bases and facilities spread across the region.

PACAF operates as the air component of the U.S. Indo-Pacific Command (USINDOPACOM), one of the eleven unified combatant commands. USINDOPACOM is responsible for military operations in the Indo-Pacific region and works closely with PACAF to ensure air and space capabilities are integrated into joint and combined operations. This relationship ensures that PACAF's efforts are aligned with overall U.S. strategic objectives in the region, thereby enhancing regional security and promoting peace and stability.

The Secretary of the Air Force and Chief of Staff of the Air Force have emphasized initiatives to prepare for future conflicts, which has caused greater interest in the workforce PACAF employs to

---

[6] Air Force Futures Requirements Oversight Team, "Capability Development Overview and Operational Capability Requirements Governance," AF/A5/7 Capability Development Guidebook Volume 2A, April 11, 2022, p. 12.

accomplish these missions. Our project sponsor was especially interested in the role technology will play in PACAF's mission and the technical expertise that will be needed. Therefore, we evaluated the supply and demand dynamics of civilian STEM talent at HQ PACAF.

As of September 2023, there were 14 civilian personnel in HQ PACAF in STEM occupations, comprising 5 percent of its civilian personnel (which totals 267 personnel). Not all these individuals had STEM degrees as their highest degree earned—four have degrees in business administration and management or law. As in other case study organizations, personnel in STEM occupations were in higher grades (grades GS-13 through GS-15) than personnel in non-STEM occupations. STEM personnel also had higher educational attainment than personnel in non-STEM occupations.

Key Insights from the Case Studies

Key insights gained from the case studies concern the supply and demand of civilian STEM capabilities, the challenges faced, and the strategies employed to address these challenges. Overall, we identified seven key insights regarding the current state of civilian STEM talent in the case study organizations:

1. Organizations prefer to hire civilians who possess general skills and operational experience over technical skills.
2. Some technical gaps exist within the civilian workforce, including skill sets in AI, data analytics, cyber, electrical engineering, radar, RF, digital modeling, and systems engineering.
3. Organizations maintain robust engagements with external partners to fill civilian technical workforce needs.
4. Organizations are unable to track technical skill sets and workforce gaps because there is no centralized database that contains detailed information about employee skills, tasks performed, and job requirements, all of which make identifying potential skill gaps within and across organizations difficult.
5. Organizations face ongoing challenges with managing, upskilling, and reskilling the civilian technical talent they do have internally.
6. The fact that the DoD civilian hiring process is slow, burdensome, and complicated introduces significant challenges to DAF units attempting to hire civilian technical talent.
7. DAF organizations are unique and distinct, reflecting different mission sets, historical organizational development, and future roles. Consequently, DAF organizations have disparate supply and demand challenges and needs for civilian technical talent.

In the following sections, we present additional supporting details from the case studies that informed each key insight. Not every key insight was mentioned in discussions with representatives from each case study organization. The insight may still apply to an organization even though it was not explicitly discussed or mentioned during discussions with that organization.

Key Insight 1: Organizations Prefer to Hire for General Skills and Operational Experience over Technical Skills

Across various programs, there is a consistent emphasis on cultivating a broad range of technical knowledge and skills within the civilian workforce, rather than seeking individuals with deep, specialized expertise. This approach prioritizes hiring motivated individuals who can be trained to meet specific programmatic needs over time and emphasizing the development of relevant competencies through on-the-job training and tailored educational initiatives.

PEO Digital Emphasizes Breadth of Technical Knowledge, Rather than Depth

Rather than hiring individuals who already possess specific technical skills, PEO Digital interviewees indicated that programs focus on developing these skills internally. This approach is based on the belief that it is more effective to hire motivated individuals and grow their skills over time to meet program-specific needs. As one interviewee noted about programs that need technical talent, "They just want a sharp person. They can grow that sharp person over time to teach them the program-specific things they need to know." Another emphasized that executing these programs is fundamentally different from what occurs in the private sector, stating, "We never have the luxury of finding someone off the street who knows. We know we have to train for and develop those competencies." Consequently, there is a greater emphasis on the breadth of technical knowledge and skills rather than depth.

Air Force Futures Wants Technologists, Not Technicians

When asked if there is a requirement for more civilian STEM in Air Force Futures, interviewees said no. One SME specified, "Not sure I need more STEM. I need maybe a broader base level of STEM understanding, but not necessarily STEM specific. I don't need STEM Ph.D.'s or Master's, but I need people with a solid foundation of STEM topics. . . . What *is* the requirement is the ability to understand why this radar system would be better in this environment than another, and not build me the radar system."

Interviewees said that they desire civilian personnel who understand the broad nature of the problem sets Air Force Futures engages and how these problems intersect with current and future technologies, a role characterized in the literature as a "technologist."[7] One SME said, "We're not a lab and not engineers from a program office. The need to understand future tech and where tech trends are going, and to understand the science behind tradeoffs, I think is critical to the holy trinity of requirements, acquisition, and resourcing."

When asked what skills they value from civilian employees, SMEs indicated they value leadership skills; strategic thinking, problem-solving, and collaboration; up-to-date academic knowledge and industry experience; and experience within the national capital region (NCR). Interviewees assessed that civilian experience, whether with industry or with other DoD entities, is an important factor for

---

[7] "Technicians develop a limited set of skills and expertise, focusing on practical knowledge in an industry or a type of technology, such as theatre or laboratory tech. Meanwhile, a technologist is an expert who specializes in technology. They possess theoretical and practical knowledge of many different types of technology, such as electronic and digital technology." Indeed Editorial Team, "Technician vs. Technologist," June 27, 2024.

Air Force Futures' efforts to understand overlapping processes as well as intersections of technology and national security.

In addition, interviewees noted that civilians serve an important function: providing programmatic continuity for Air Force Futures. One SME noted that civilians "already built out relationships and have seen several attempts that have gone up and failed. As we get into details of the capability development plans, we will rely on the continuity of the understanding as well."

Headquarters, Pacific Air Forces Prefers General Skills and Operational Experience over Specific Technical Skills

PACAF values flexibility and the ability to learn over specific technical skills. This is driven by the command's operational nature, which demands adaptability. As one interviewee stated, "We assume everyone we get can be flexible enough to learn new things. We look for people who are eager to learn." Another interviewee highlighted the operational focus: "The most critical skill set is to understand what the ops world needs." A third perspective reinforced this view: "We need multidisciplinary people who can think. . . . It's useful to have people with STEM backgrounds but out here its operational and we're not doing things people in traditional STEM do." Operational experience is prioritized over technical knowledge, with an emphasis on applying technology in operational contexts rather than understanding the underlying science.

These preferences shape hiring practices and the types of skills valued within the organization. Examples include the following:

- Positions requiring knowledge and application of underlying science are often outsourced.
- There is a reliance on occupational series requirements as the sole representation of the KSAs needed for the position. As one interviewee noted, "There are a lot of things we look for [when we are hiring], but we are beholden to what the job series is defined as."
- Since hiring emphasizes general skills, there is an acceptance that technical expertise will need to be developed through on-the-job training and training courses.

Key Insight 2: Organizations Face Unique Technical Gaps and Emphasized the Need for Operational Understanding

Except for cyber, gaps in technical skills were generally unique to each case study organization. Other gaps suggested a need for technical personnel to better understand operational mission sets.

PEO Digital Identified Specific Needs for Skills in Engineering, Radar, and Digital Modeling

PEO Digital interviewees indicated that programs are generally stable, and they do not expect completely new demands in the near future. Despite this stability, some gaps in technical talent have been identified. Hiring and retaining cyber professionals have proven to be more challenging. PEO

Digital also has specific needs for skills in electrical engineering, radar, RF, digital modeling, and systems engineering.[8]

Headquarters, Pacific Air Forces Identified Gaps in Artificial Intelligence, Data Analytics, and Cyber

Discussions with HQ PACAF personnel revealed some gaps in technical competencies. First, there is a broad recognition that directorates should be developing expertise in AI and data analytics methods. Although no specific mission-essential tasks currently employ AI or data analytics, there is a prevailing sentiment that these technologies should be leveraged for operational, functional, and administrative tasks. One initiative addressing AI and data analytics gaps is Project Phalanx, spearheaded by the Secretariat of the Air Force Studies and Analysis (SAF/SA) in response to an Air Force Chief of Staff directive.[9] This initiative aims to strengthen and establish independent operations research analytic organizations within MAJCOMs to ensure decision dominance and improve warfighting capabilities. Over the next four years, Project Phalanx will provide approximately 20 positions to HQ PACAF to incorporate data science techniques and products into operational activities. This effort is expected to provide the PACAF commander and the commander of USINDOPACOM with improved information for decisionmaking.

A second identified gap was for the hiring and retention of cyber talent. While this was recognized as a gap, several interviewees praised the hiring and retention processes under the Cyber Excepted Service (CES) as much better than traditional hiring and retention programs and practices.[10] This is another case where HQ PACAF relies on external organizations for the provision of its workforce: Headquarters Air Combat Command is responsible for CES hiring at HQ PACAF.

The availability of personnel overall, including persistent vacancies and frequent turnover, was of more concern to interviewees than the acquisition of civilian employees with specific technical competencies. One individual mentioned that if they had more resources, they would not hire additional civilian employees, but rather they would hire more contractors to "take away day-to-day tasks and minutia so that civilians and active duty can work on bigger thought pieces."

There was also little appetite for attempting to hire more civilians with very technical academic degrees (e.g., a Ph.D. in engineering or physics). One interviewee clearly stated that "experience is more important than a degree." Another interviewee emphasized that experience coupled with other

---

[8] Because systems engineering is not a career field and engineering departments typically do not offer systems engineering as a degree, the DAF will take other engineers (e.g., mechanical engineers) and assign them to systems engineering tasks.

[9] Jasmine Braswell, "Det 4 Provides a Data Analysis Capability for ACC Units," Air Combat Command Public Affairs, October 12, 2023.

[10] CES provides DoD with enhanced authorities to offer competitive salaries and benefits that are more aligned with private-sector standards, thus making cyber positions more attractive to top talent. These flexibilities include the ability to expedite the hiring process, which allows DoD to quickly onboard highly skilled cyber professionals who are in high demand. Additionally, CES permits the use of recruitment and retention incentives such as signing bonuses, student loan repayments, and special pay rates to retain critical personnel. The service also offers career development opportunities and clear advancement pathways, which are essential for maintaining a motivated and proficient cyber workforce. By leveraging these flexibilities, DoD aims to build a robust and resilient cyber defense team capable of addressing the evolving threats in the cyber domain. For additional information, see David Knapp, Sina Beaghley, Troy D. Smith, Molly F. McIntosh, Karen Schwindt, Norah Griffin, Daniel Schwam, and Hanna Hoover, *DoD Cyber Excepted Service Labor Market Analysis and Options for Use of Compensation Flexibilities*, RAND Corporation, 2021.

types of certifications such as Project Management Professional may be of more practical use in certain functions such as information technology than very technical degrees.

Of note, not all PACAF interviewees identified gaps. One representative from PACAF/A2 stated, "I would not tell you that I would add more STEM folks" and cited the directorate's partnership with external organizations such as the National Geospatial-Intelligence Agency and the MITRE Corporation as mechanisms to fulfill technical capabilities. The interviewee echoed others' comments that the most critical gap is in civilian personnel with experience in science and technology who can make connections to operational mission sets. According to one interviewee, "The most critical skill set is to understand what the operations world needs. . . . We want operations personnel to have STEM background, and science and technology people to have operations experience. That's the perfect world."

Key Insight 3: Organizations Leverage External Partnerships to Address Technical Skill Gaps

Interview discussions suggested that the civilian workforce was only one factor to consider when making decisions on how to address gaps in technical capabilities. Specifically, SMEs indicated that technical needs can also be addressed using a combination of contractors and uniformed and federally funded research and development center (FFRDC) personnel. In addition, other organizations across the DAF may have personnel with technical expertise who can provide temporary support to fill gaps.

PEO Digital Uses Contractors Extensively, with Benefits and Drawbacks

PEO Digital contracts with a variety of external organizations to obtain technical expertise. Examples include MITRE, Georgia Tech Research Institute, Massachusetts Institute of Technology (MIT) Lincoln Laboratory, and the U.S. Department of Transportation's Volpe Center. Participants indicated that PEO Digital uses such contractors extensively. One participant said that PEO Digital relies "extremely heavily" on contractors, and another said there were twice as many contractors as civilian engineers in their organization. A third participant said that PEO Digital's use of contractors differentiates it from other Air Force organizations and that "some other directorates are totally government civilians and don't look to contractors. We are the balanced approach. If we can't get the right civilian, we'll look somewhere else."

Contracting with FFRDCs also enables PEO Digital to obtain deep expertise needed on a temporary basis or at certain points in the life cycle of a system. These experts can provide the technical expertise needed to oversee other contractors such as Lockheed Martin or Northrup Grumman, that supply systems to the Air Force; help move systems into the field quickly; and help fill gaps that result from unfilled positions or lack of experience among PEO Digital engineers:

> MITRE and Lincoln Labs are incredibly talented. I use them for challenging issues. The radar SME has been working in the field for 30 years, so when Lockheed Martin presents their systems, they give an objective evaluation versus a military officer who's been there one year. . . . It's great that the officer can understand the requirements from a system engineering perspective, but in terms of hard technical problems, I rely on the FFRDC.

During our discussions participants expressed mixed opinions about PEO Digital's use of contractors. Some described it as largely positive and said there were no major gaps in the civilian workforce. However, others said they would prefer that some functions performed by contractors were instead performed by PEO Digital employees, but that PEO Digital lacked enough funded positions or filled billets to carry out those functions. Several participants mentioned the high cost of using contractors, and one described a specific contractor as "good at pointing out things but not good at getting solutions."

Air Force Futures Leverages Contractors and External Partners for Technical Skills

RAND project members asked Air Force Futures SMEs how they fill gaps in civilian technical knowledge when they have requirements for STEM capabilities that their workforce lacks. Interviewees remarked that they leverage a range of partners to fill these gaps. Partner engagements include working with other divisions in the directorate, partnering with other HAF directorates, engaging with personnel in AFRL, utilizing contractors, employing Total Force augmentees (i.e., National Guard or reserve military personnel), or reaching outside DoD to private industry organizations. SMEs also indicated that they rely on their own civilians to identify points of contact in other organizations for accessing data and information on technical requirements; this indicates that Air Force Futures uses civilians in a networking and linking role across organizations.

Engaging external partners to fill gaps in civilian technical talent can lead to beneficial opportunities for building relationships. One interviewee said that

> we do a lot of industry partnership. An example of that is my artificial intelligence lead brought in Amazon leadership two weeks ago because they have security clearances but never had a TS [Top Secret] level discussion. We brought them in and gave them a TS view of adversary threat and in turn that better helps them understand where we are coming from in the DoD, so that helps us bridge that gap.

Alternatively, relying on external workforces to fulfill technical knowledge gaps introduces challenges alongside benefits. One Air Force Futures SME said,

> I will say the civilian gaps have been filled with contractors. The challenge is context. It takes a while to spin them up and get them to the level we need them to perform. I would say it slows output significantly when we augment with temporary workers or borrowed workers. [As a positive], they bring a fresh perspective to the team. We have so many people who have only worked with the Air Force—the contractors bring a fresh perspective. When we're looking at a skill set that's perishable, their upskilling seems to be better than ours. We have someone who got an M.B.A. 30 years ago and they may not have the skills our contractors have.

Headquarters, Pacific Air Forces Relies on External and Reach-Back Organizations for Technical Support

HQ PACAF directorates rely heavily on external and reach-back organizations for technical support. Interviewees cited examples such as Air Combat Command, FFRDCs, AFRL, the National Air and Space Intelligence Center, the National Geospatial-Intelligence Agency, and multiple contract vehicles administered by HQ PACAF or Air Force entities. The availability of these external organizations allows PACAF to leverage specialized expertise without having to maintain a large in-house staff for every technical domain.

HQ PACAF recognizes its role as a component of USINDOPACOM, which drives capability requirements for PACAF and emphasizes operational readiness for a "fight tonight" approach. This focus on current operations means that identifying future technical skill requirements is delegated largely to individual DAF functional managers. For instance, PACAF/A2 depends on the DAF-level functional manager to determine the evolution of training for the intelligence series (0132) in terms of additional KSAs and required training.

Key Insight 4: Department of the Air Force Organizations Face Talent Management Challenges for Technical Civilian Workforce

The DAF faces significant challenges in tracking and managing technical skills across its workforce.[11] Currently, there is no centralized system for identifying the current and future demands and supply of technical skills. This results in an incomplete understanding of workforce needs. These challenges can be attributed, in part, to specific technical skills that are managed by lower levels within the organizations. In other cases, needs may not be communicated when there is a lack of funding that contributes to perceptions that additional efforts on workforce planning will be ineffective (i.e., requests for additional personnel will not be supported).

PEO Digital Lacks Systems to Track Technical Skills

Interviews with PEO Digital SMEs reveal that the DAF currently lacks a centralized system for identifying both current and future demands and supplies of technical skills. Instead, technical skills are managed within individual directorates and their subordinate levels, such as divisions and branches. Workforce requirements are often determined informally through a process referred to as "mapping out the work."

Recognizing the limitations of this approach, some efforts aimed at improving the tracking and development of technical skills have been underway or are currently being planned in PEO Digital. These include the establishment of a new talent acquisition function and the development of a competency taxonomy that spans from broad domains, such as electronic warfare, to specific skills, such as signal modulation and signal filtering. However, it is important to note that previous competency frameworks have not been used consistently across programs. Some programs indicated not using them at all, whereas others indicated that they are used only to inform the hiring and internal staffing of open positions. Even then, the usefulness rests heavily on the individual competency managers who may not have sufficient understanding of the position requirements.

---

[11] *Talent management* has a wide range of definitions. For the purpose of this report, we adopt OPM's definition: "A system that promotes a high-performing workforce, identifies and closes skills gaps, and implements and maintains programs to attract, acquire, develop, promote, and retain quality and diverse talent." OPM, "Talent Management," webpage, undated-c.

Air Force Futures Uses In-House, Experiential Learning for Technical Skills Training

During interviews, we asked Air Force Futures representatives about opportunities to reskill or upskill their technical civilian workforce. Overall, SMEs indicated that they focus such investment on experiential learning or on learning from those singular individuals in Air Force Futures who possess advanced technical skills. One SME noted that technical professional development resources exist but can be difficult to access: "I know every year the civilian personnel community does a call for courses. The way that is advertised is they give a list of 40 different websites you can sift through and try to find different courses relevant to your job series, and it's not very helpful. We know there is training out there but getting access and being aware is the challenge for us."

Another SME mentioned rare partnership opportunities with academic organizations as a mechanism to invest in maintaining and growing civilian technical talent: "There are some courses that are set that we can send people to. MIT is one, Stanford is another. There are some of those opportunities, but those are few and far between. Normally when they [civilians] get to the Pentagon, the development opportunities are rare."

Finally, one interviewee identified the role of in-house training for upskilling and reskilling civilian technical talent:

> There's in-house training we developed [which] I think is helpful for people who haven't been in our environment before. We try to have a flat organization, so the ability across our teams to link up the right people to have the right conversations to get after these capabilities to trying to force design becomes easier the more you do it, and because of our onboarding process. There are some formal and informal courses we take.

Headquarters, Pacific Air Forces Lacks Centralized Planning for Technical Requirements Across Divisions

PACAF/A1 is responsible for human resource functions within HQ PACAF, including manpower authorizations for civilian positions as well as the hiring of individuals into these positions. Discussions with PACAF/A1 revealed that they respond primarily to needs determined independently by each directorate, rather that assessing STEM civilian talent requirements across the headquarters. Each directorate's unique operational responsibilities necessitate specific competencies, resulting in significant differences in technical skill requirements. For example, PACAF/A2 requires data analytics skills for intelligence analysis, PACAF/A9 seeks programming and modeling expertise coupled with operational experience, and PACAF/A3 focuses on technological awareness for operational applications.

Requirements for specific competencies are addressed by the individual directorate or work center, or in some cases requirements are communicated to DAF-level functional managers, but there is no cross-division aggregation of technical skills or information on skill gaps and vacancies. In addition, there is no centralized planning related to workforce mix decisions—that is, whether a position should be held by an officer, enlisted member, or civilian employee. Hiring managers in each directorate and work center tend to operate independently, and this can potentially lead to an incomplete understanding of technical needs across the organization. As the DAF prepares for great power competition, one interviewee from PACAF/A1 stated that "we need to do a mission analysis [of] the actual gaps" in order to determine the need for technical personnel in PACAF.

Key Insight 5: Department of the Air Force Organizational Structures Complicate Filling Specific Technical Skills Gaps

Addressing gaps in technical talent can be done by authorizing and funding new billets. However, the reality is that the DAF has a constrained budget and some programs are operating at levels well below what is required. Competing priorities also place considerable pressure on programs to execute their missions. Gaps can also be addressed through training and education; however, awareness of what is available and whether available training meets organizational needs is unclear.

PEO Digital Faces Challenges in Securing and Funding Billets

While the DAF has been able to hire sufficient talent, there are broader concerns about consistent gaps in funding requirements and securing the necessary billets for their programs. As one PEO Digital interviewee noted, securing billets remains "infinitely more difficult" than finding and hiring the right talent. Several other challenges were noted, including programs that are funded significantly below their full manpower requirements. For example, the program with the highest allocation is operating at only 60 percent of its needed manpower. Without sufficient funding, older systems are struggling to maintain basic operations, which further contributes to insufficient support for modernization efforts. Finally, division chiefs may report to multiple program executive officers, leading to competing priorities. Divisions may also receive additional mission areas without corresponding increases in personnel, thereby further straining resources.

Headquarters, Pacific Air Forces Faces Challenges in Balancing Technical and Operational Billets

Interviewees indicated that the demand for technical competencies is tempered by broader organizational challenges within the directorates. In the DAF, manpower resourcing comes from MAJCOM, which makes HQ PACAF responsible for balancing the need for additional technical personnel against other manpower requirements, including operational positions. Unfunded manpower authorizations and staffing shortfalls are common across all HQ PACAF functions, so STEM needs may or may not be identified and communicated due to the low likelihood of additional manpower authorizations being funded. When asked about requirements for additional STEM positions, one interviewee commented, "PACAF has unfunded requirements, so initially I anticipate this will result in more unfunded requirements. Further, division and branch chiefs are focused on their immediate responsibilities, leaving little time for strategic workforce planning and identifying future STEM needs."

Key Insight 6: Department of the Air Force Organizations Face Process-Driven Civilian Hiring Challenges

Interviewees raised multiple concerns about barriers to navigating federal civilian hiring policies and practices. These barriers included slow hiring timelines, lack of competitive pay, lack of flexibility to make changes to PDs without creating additional delays, and lack of training for hiring managers.

Air Force Futures Identified Civilian Hiring Challenges as a Significant Constraint

A key theme that emerged across all Air Force Futures interviews is the challenges associated with hiring civilians, whether for STEM talent or more generally. Interviewees articulated challenges with defining and updating requirements in PDs to gain the right capability; the slow hiring process at the DoD level; and low pay in the NCR compared with private industry.

Position Description Challenges

Interviewees identified the challenges associated with defining the requirements for hiring civilian technical talent. One SME described this challenge as "the yin yang of civilian hiring: You don't want a position description that's so specific you narrow the field of candidates or so wide you have an infinite number of potentials. There's a sweet spot."

Organizationally, changing PDs as a mechanism to attract talent or new skill sets is burdensome and complicated. One interviewee noted that

> I inherited the position descriptions and billets. We have modified them somewhat to align them better. Ultimately, I'm responsible for rewriting the PDs if necessary. I'm hesitant to do that because it's not an easy process and it takes a long time and delays hiring new people. A lot of time the PD has little or no bearing on the skills of the people we're hiring so I don't do a lot of tinkering around with PDs.

Another interviewee echoed this hesitancy, saying

> I think what we need for an organization is to tweak the position to get this skill set to morph to where we see the organization is going. That requires you to change the position description and the classification of the position description, which is what takes the bulk of time. I'm hesitant to change a position description which goes against the need to change as the organization. I can get the body quicker [without changing the PD], but it may not be the body I need to actually make change in the organization.

Time Challenges

Compounding the challenges associated with updating PDs to reflect emerging technical requirements, civilian hiring processes are burdensome and lengthy due to DoD process requirements. For example, one Air Force Futures interviewee said that if they could fix any challenge associated with civilian technical talent, they would accelerate the civilian hiring process "because it's broken to a point that it's affecting mission. It's difficult to hire anybody. It takes a long time. It's so long in the past we've hired people and because it takes so long to get them on our books, they get other jobs. It takes 8–12 months to hire a civilian employee."

One specific time challenge associated with civilian hiring is the security clearance process, which takes so long that hired employees with desired STEM backgrounds decide not to wait on the clearance process to complete and instead move on to other opportunities.

Pay Challenges

An additional challenge in hiring civilians is pay disparity with the private sector, particularly when hiring technical talent. One Air Force Futures SME said that

> if you want technical knowledge of a specific type—for example artificial intelligence— it will be difficult for the military to offer compensation as much as the industry can provide, meaning we won't get top-tier talent. We have patriots and other people who want to serve our country, but it does drastically limit our pool.

Air Force Futures Mitigation Strategies

One interviewee noted that to mitigate the challenges associated with hiring new talent for technical roles, "We recycle the same people from department to department to department." Another interviewee noted that DAF hiring processes lack familiarity with hiring external talent into higher-ranking positions, such as "a GS-13, 14, or 15. I've overheard a couple of complaints about the personnel offices not knowing how to adapt to that. Everyone is so used to hiring internally."

Some processes exist to attract talent despite these hiring challenges. For example, one interviewee noted that with their office's last open civilian billet, "We've been able to go external and offer it as a remote position. That allowed us access to more talent: we had 800 applicants. We had 30 applicants that were highly qualified." Yet even with this flexible approach, the office still took eight months to fill the position with a new hire.

One Air Force Futures SME recommended establishing a larger hiring pipeline for civilian technical talent at lower grade levels and based outside the expensive NCR. "If there was a way to build outposts somewhere else, we can build that training pipeline of technical experts so we can grow our own pool of individuals who understand how the DoD works and provide more insight than the people who live most of their lives in D.C. proper. That could help."

Headquarters, Pacific Air Forces Identified Location-Specific Challenges in Hiring and Retaining Technical Talent

Interviewees highlighted significant challenges the DAF faces in hiring and retaining civilian personnel, particularly in technical fields. Structural constraints within federal recruiting and hiring practices—such as lengthy processes and bureaucratic hurdles—significantly affect HQ PACAF's ability to attract and retain technical talent. The small civilian talent pool, along with difficulties retaining civilians due to limited advancement opportunities and competition from the private sector, can exacerbate these challenges.

Overseas Location Challenges

At HQ PACAF, hiring challenges are further compounded by the high cost of living and other location-specific issues in Hawaii. Hiring managers often feel compelled to adjust occupational series and omit specific needed skills to fill vacancies, recruiting based on availability rather than actual needs. The reported workforce gaps appear to be a function of multiple factors, including unfunded requirements, limited local labor supply, and competition for the same talent. Local government salaries have not kept pace with the cost of living, and contractors may offer higher salaries. Addressing basic hiring needs is a bigger concern than hiring for specific technical knowledge.

Lack of Training for Hiring Managers

The bureaucratic processes involved in hiring and classifying positions pose significant barriers to acquiring the right skills. The time and effort required to fill civilian vacancies are burdensome and often result in losing candidates to other opportunities. As one interviewee noted, "The time and effort to fill civilian vacancies is taxing, especially given other high-priority responsibilities." Hiring managers mentioned not being trained in the administration required to post a position, recruit potential employees, and evaluate and interview candidates. When discussing taking on a hiring official role in HQ PACAF, one interviewee remarked, "I was not well trained to do interviews or recruiting. I had to do my own research in best practices. The guidance from Air Force is hodgepodge, not well displayed or presented, and not at all useful." Additionally, several HQ PACAF representatives focused on PDs as a particularly burdensome part of the process. As one stated, "The PD itself is overly complex and hasn't been modernized in decades. . . . There is nothing easy about it." Ultimately, interviewees shared that they manage the best they can with the difficult system available to them: "We work within the system we are given."

Key Insight 7: Diversity of Department of the Air Force Organizations and Mission Sets Produces Diverse Needs for Civilian Technical Talent

Interviewees noted the specific nature of Air Force Futures when providing insights on civilian needs and hiring. Specifically, interviewees framed the demand and supply dynamics of civilian technical talent within the organizational history, mission, and transformation of Air Force Futures. One interviewee concluded that

> I think the one thing I would leave with you is we are a different type of organization than most in the Air Force. We have the ability to interact within the organization and collaborate outside of the organization. The one thing we do have in common is our civilian hiring process, which at all levels tries to get us the bodies we need in the short amount of time but satisfices the specific need for speed and doesn't deliver on either side of that equation. So [the civilian hiring process] doesn't deliver in quality or speed, so neither one gets satisfied.

HQ PACAF, in contrast, faces a nuanced landscape in balancing the need for technical competencies with broader organizational challenges and structural constraints, all within a unique overseas operational environment. The autonomy of directorates in determining skill requirements, coupled with a preference for general skills and operational experience, shapes the hiring practices within the organization. Reliance on external and reach-back organizations provides a crucial supplement to in-house capabilities, yet challenges in hiring and retaining technical talent persist, exacerbated by bureaucratic hurdles and location-specific issues. While initiatives such as Project Phalanx aim to address gaps in AI and data analytics, and while CES offers a more efficient hiring process for cyber talent, the overall struggle to attract and retain the right personnel remains a concern. Ultimately, addressing these challenges will require efforts to streamline hiring processes, enhance strategic workforce planning, and leverage both internal and external resources effectively.

Overall, civilian hiring practices are complex, as shown in previous RAND research.[12] Indeed, one Air Force Futures SME shared that hiring practices are complicated by competing priorities set at the Air Force's Personnel Center (AFPC):

> We rely on AFPC to send us lists of eligibles based on their mostly by-law requirements for priorities and eligibilities. We can't just find a great candidate and hire them. We have to have them apply for positions through USAJOBS, and AFPC gets to decide if they are forwarded to us for consideration. In the past, we have aggressively recruited some individuals only for AFPC to determine they either are not eligible for some reason or that they get out-prioritized by all the priorities that have been established (Priority Placement Program, Military Spouse, Veteran's Preference, Disabilities, etc.).

These SME contributions point to our final key insight: Despite facing the same civilian hiring constraints as occur within the DoD system, DAF organizations are unique and distinct due to their mission sets and organizational history. This history shapes civilian billet structures. Not all DAF units may require more civilian technical talent. In addition, most DAF units work with what they have to get the mission done. Indeed, identifying requirements and spending limited time or manpower to correct billet structure runs counter to the DAF culture of "make it work with what you have."

Conclusion

Several comprehensive themes emerged from our discussions with case studies representatives. First, DAF organizations have unique needs for civilian technical talent. These needs vary depending on unit mission, existing workforce make-up, and future mission requirements (which may change as the DAF prepares for future conflicts). Second, DAF organizations face significant challenges in hiring new civilian personnel because of the burdensome requirements of DoD's civilian personnel system. Third, because of these challenges, organizations have adopted dynamic approaches to addressing gaps in civilian technical capabilities when they do exist. These approaches include leveraging external organizations to fill technical gaps and training current workforce members on emerging technical skills, when possible.

While findings from these SME discussions are specific to each case study, overall the themes point toward the DAF facing enduring, systemic challenges in its civilian technical workforce. These challenges involve both the civilian hiring process and the lack of a centralized mechanism within and across organizations to identify and track technical gaps in the workforce. Considering this shortfall, we consider surveys as a tool to aid organizations in identifying and documenting technical needs within the civilian workforce in our next chapter.

---

[12] Keller et al., 2023; Groeber et al., 2021.

Chapter 5

Using Surveys for Supply and Demand Assessments of Technical Skills

As reflected in the literature and in our case study approach described in previous chapters, there does not appear to be a comprehensive, competency-based mapping of the skills the DAF requires among its civilian STEM-focused personnel. Because job descriptions themselves are tied to a specified occupational series, as opposed to a more narrowly articulated set of technical knowledge or skills, there are inherent limitations in how needs can be assessed and understood. Other tools such as job analysis worksheets, which may capture more specific work activities and requirements, do not appear to be regularly used or updated. And while interviews such as those discussed in the previous chapter yield nuanced information from within different organizations, they are a time-intensive approach to generating supply and demand data.

Given these challenges, we considered the kinds of questions that might help work unit supervisors, and potentially even higher management levels, understand and define their current and future needs for STEM capabilities. We aimed to consider these questions in a way that was not restricted by occupational series labels but that incorporated what they, in many cases, are ultimately seeking to capture: the specific competencies needed for the various jobs tasked to government civilians.[1]

Toward this end, we developed two survey options to gather information regarding civilian workforce supply and demand from different organizations. First, we developed a comprehensive survey through discussions with PEO Digital SMEs. Second, we conducted an informal, pared-down survey with Air Force Futures SMEs. We discuss both surveys' design and execution below, along with recommendations for further testing and using these survey options to monitor current and future organizational needs.

Designing a Survey: Supply and Demand for Technical Competencies

To design an assessment of civilian STEM personnel competency needs, we developed a survey through multiple iterations of research, stakeholder feedback from PEO Digital SMEs, and revision. As a starting point, we reviewed standard personnel assessment questions about competencies, work

---

[1] Of note, we considered existing survey structures such as the Defense Competency Assessment Tool (DCAT) managed by the Defense Civilian Personnel Advisory Services. The primary challenge to using DCAT is that competencies first need to be developed and defined. As mentioned previously, the effectiveness of previous efforts to implement competency frameworks has been mixed. There is no current set of competencies that covers the range of occupational specialties in our case studies. Another challenge is that DCAT requires supervisors to evaluate the proficiency of every employee. Although results can be aggregated to branch, division, or higher levels, the resources required to execute effectively are extensive. See Defense Civilian Personnel Advisory Services, "Competency Management," webpage, undated.

activities, and qualifications provided by the O*NET Resource Center website and OPM's website.[2] Using these resources, along with OPM's *Handbook of Occupational Groups and Families*—which codifies STEM-related government positions—we developed an initial set of questions for unit supervisors.[3] The survey items aim to identify the specific competencies required in various work units and to determine organizational needs for STEM civilians.

The first iteration of the survey, created using an Excel spreadsheet, focused primarily on two issues: the number of personnel in a specific occupational area *currently* working in the unit and the number that would be *preferred*. This early version was designed to capture proficiency levels and asked respondents about the education levels of their employees in each relevant occupational area— bachelor's, master's, or doctoral degrees.

Next, we included additional categories based on internal RAND feedback to address other relevant items:

- the number of positions *authorized* for a specific knowledge area or technical competency (not just *currently* filled positions and the *preferred* number of positions)
- the occupational series under which the knowledge area or competency could be included[4]
- current work unit proficiency gaps for the given knowledge area or competency
- the degree of future risk to the work unit's proficiency in the given knowledge area or competency
- the preferred approach for developing essential competencies, whether through outside hiring, on-the-job training, coursework, or other means
- hiring challenges faced by the work unit, such as private-sector wage disparities, excessive time-to-hire, and undesirable work locations.

Finally, we gathered feedback from the PEO Digital case study supervisors, including division and branch chiefs, to assess how the questions were interpreted and whether responses provided the intended data we needed. The full survey is reproduced in Appendix D.

PEO Digital Stakeholder Feedback

Several insights emerged from multiple conversations with PEO Digital division and branch chief engineers who reviewed or completed the STEM needs assessment survey. One key insight was that a comprehensive competency survey would likely be too detailed. For instance, the distinction between having a master's degree or a doctoral degree may be less relevant than the number of years of experience an individual has in a given field. This feedback suggested that while experience is valuable, determining the appropriate level of detail may be challenging.

Another insight was that competency tracking largely ceases once individuals have joined a work unit. While specific competencies may be evaluated during hiring to address a specific, immediate

---

[2] O*NET Resource Center, "O*NET® 28.3 Database," webpage, undated; OPM Management, "Competencies," webpage, undated-a.

[3] OPM, 2018.

[4] We preloaded the survey with knowledge areas and competencies based on previous discussions with SMEs from PEO Digital. However, branch chiefs were directed to add other technical knowledge areas that were critical to their mission.

need, detailed competencies and proficiencies are not updated or consistently maintained across work units for long-term tracking.

In addition, responses to the STEM competency needs survey were inconsistent across categories that might be expected to correlate. For example, a multiple-person shortage in a specific competency might be expected to positively relate to both a current proficiency gap *and* a future risk to work unit proficiency. However, responses across these categories did not align in this way. That is, a shortage of personnel in a particular competency did not necessarily equate with a current or future proficiency risk in that competency. This insight revealed that there is not only an absence of a systematic, current assessment of personnel competencies, but also an absence of a uniform understanding of the relationship between current and authorized personnel numbers, proficiency among current personnel, and both current and future proficiency gaps.

Taken together, stakeholder feedback strongly suggested that comprehensive and current documentation of required competencies for DAF STEM-focused civilians is largely absent and generally challenging to implement. Most hiring solutions are addressed locally rather than at the MAJCOM or department level. Stakeholders generally reinforced key insights gleaned from interviews and indicated that hiring needs could be met through external hiring, contract or FFRDC support, or military personnel support and that the absence of a systematic approach to competency management for STEM civilian personnel was not viewed as a glaring weakness in work unit operations.

However, these responses also highlighted that, without a systematic approach, awareness of personnel competencies becomes less clear at higher organizational levels. Without a centralized system to store relevant information about the technical knowledge and skills of DAF civilians, addressing personnel needs will likely remain a highly localized process. This gap affects the DAF's ability to quickly address challenges by placing qualified personnel where they are most needed to meet current and future demands.

PEO Digital Survey Key Insights

In summary, our efforts to develop, test, and refine a survey on STEM competencies among DAF civilians revealed several interconnected issues. First, consistent with previous research, hiring challenges—such as higher private-sector wages, lengthy hiring processes, and structural limitations focusing on occupational series rather than specific skills—continue to affect the DAF. Second, because of the interconnected nature of issues such as personnel numbers, competencies, proficiency gaps, and hiring challenges, a stand-alone survey without accompanying discussions would likely be insufficient to systematically identify competency needs. Using a survey to identify competency requirements, potential gaps, and risks to missions should be considered an important but not a sufficient step in prioritizing areas to target for talent management programs. Once survey responses are completed, follow-on discussions with branch and division-level supervisors are needed to more clearly understand why some gaps identified by the survey were not viewed as current or future mission risks by supervisors. Holding these follow-on conversations would be an important step to help the DAF know where to concentrate efforts and resources to improve mission effectiveness.

While developing and refining the survey highlighted the difficulty of mapping competencies, it also underscored the potential value of increasing the awareness of the competencies held by the civilian workforce—particularly for supervisors, hiring managers, and personnel directors, especially in technical skills. We explored this awareness in more detail through a second survey targeted to work units that may have less well-defined needs for technical talent.

A Second Survey: Valuing Characteristics for Current and Future Needs

Comprehensive surveys with lists of technical competencies are an important tool for identifying technical workforce needs and gaps. However, smaller organizations that have mission sets less clearly tied to technical requirements may benefit from an alternate survey design. Therefore, we explored a second option as a mechanism to establish current and future needs for civilian technical talent: administering a simple, informal assessment of the characteristics that an organization most values within its civilian workforce. Toward this goal, we asked the six interviewees from Air Force Futures the following question: *When considering the current and future needs of Air Force Futures, what characteristics in a civilian hire would be more important to help Air Force Futures meet mission objectives?* From those SMEs we contacted with this informal survey, five completed the survey and one participant did not respond.

Survey participants were asked to consider eight characteristics and indicate how much they value the characteristics in a civilian hire to meet Air Force Futures' *current* needs and to meet *future* needs. Participants weighted characteristics on a scale from 1 to 5, with a value of 1 indicating that the characteristic has a low value and 5 indicating that the characteristic has a high value. The survey also provided opportunities for participants to generally describe their ideal candidate for hiring a new civilian into Air Force Futures.

The eight characteristics are as follows:

- academic degree—bachelor's with STEM focus
- academic degree—bachelor's with non-STEM focus
- academic degree—master's or Ph.D. with STEM focus
- academic degree—master's or Ph.D. with non-STEM focus
- technical expertise (i.e., expertise in AI, modeling, data analysis, and so on)
- DoD experience (i.e., work experience in a DoD organization)
- private-sector experience (i.e., work experience in private industry)
- NCR experience (i.e., work experience in NCR).[5]

Appendix D contains the full survey.

The survey test had a small sample of five respondents. Therefore, the results presented below are not conclusive and should be viewed as a preliminary suggestion for how the data could be analyzed and depicted to provide useful insights into current and future STEM demands. As shown in Figure 5.1,

---

[5] When someone says they need to hire someone with "NCR experience," it typically means they are looking for an individual who is familiar with the political, bureaucratic, and operational environment of Washington, D.C. and more specifically service and DoD headquarters organizations in the Pentagon and surrounding area. This includes understanding how federal agencies, Congress, and defense contractors interact, as well as having experience navigating the complexities of government processes and decisionmaking.

survey participants value academic degrees with a STEM focus (master's or Ph.D. and bachelor's), technical expertise, and private-sector experience. Participants indicated that they valued these characteristics in supporting Air Force Futures' current operational needs as well as the organization's engagement with future challenges. In providing an explanation for their value ratings, one participant who valued *academic degree—bachelor's with STEM focus* explained that "STEM degrees tend to have a higher level of logic and quantifiable answers. They also tend to be less subjective and less emotive." Another survey participant who valued *academic degree—master's or Ph.D. with STEM focus* for both current and future requirements linked this characteristic to technical knowledge with the comment "I would be excited to hire an applicant with this background as long as they were able to talk through the integration of new technology into the force in a more rapid manner."

Figure 5.1. Air Force Futures Survey Results, Averages

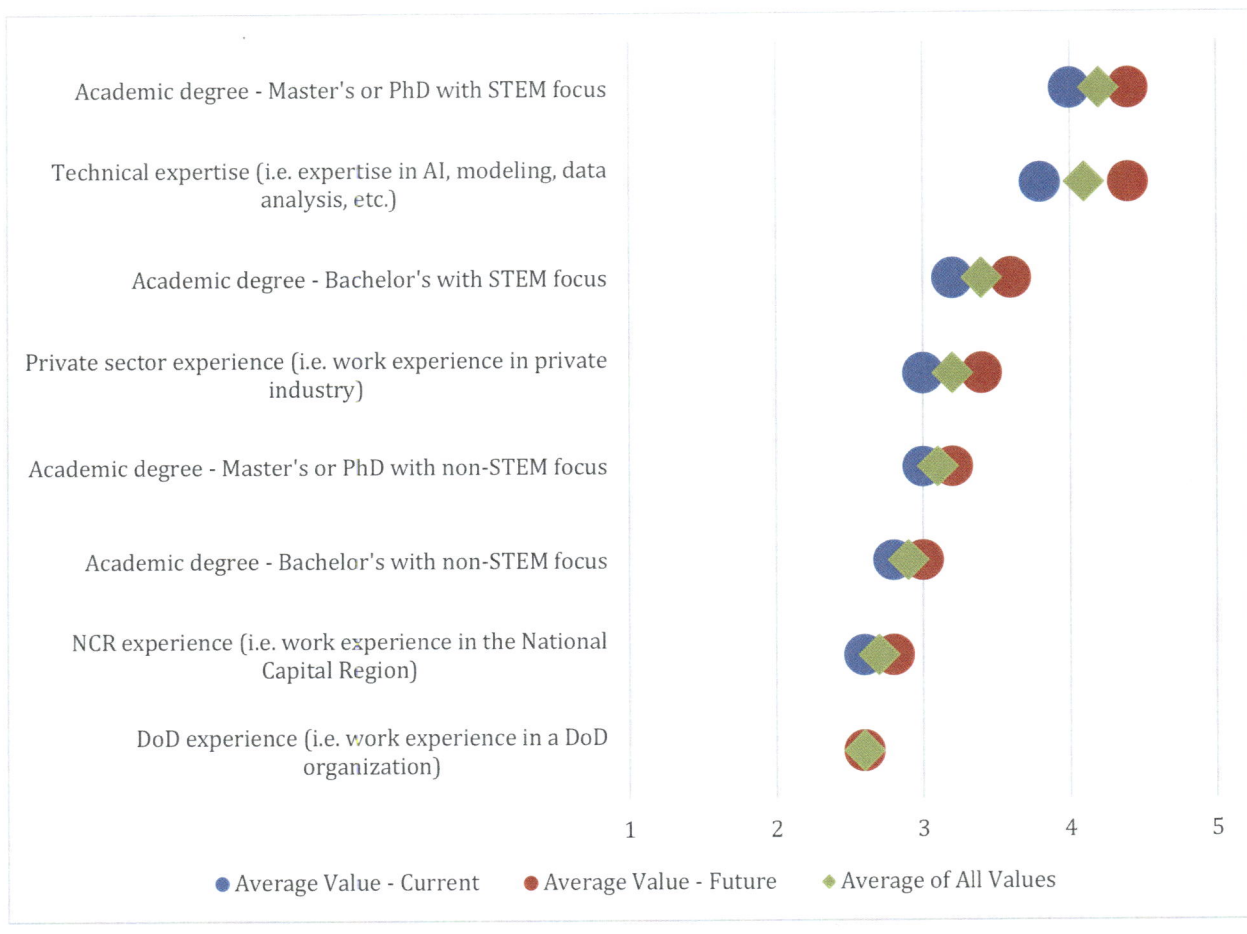

Figure 5.2 depicts a radar graph comparing average current and future values for these characteristics. This visual method shows that the five respondents did not indicate significant difference between current and future needs. The largest change from current to future values lies in

the characteristic of technical expertise, with the current value average across participants of 3.8 and the future value average of 4.4 (a change of 0.6).

Figure 5.2. Air Force Futures Survey Results, Current and Future Changes

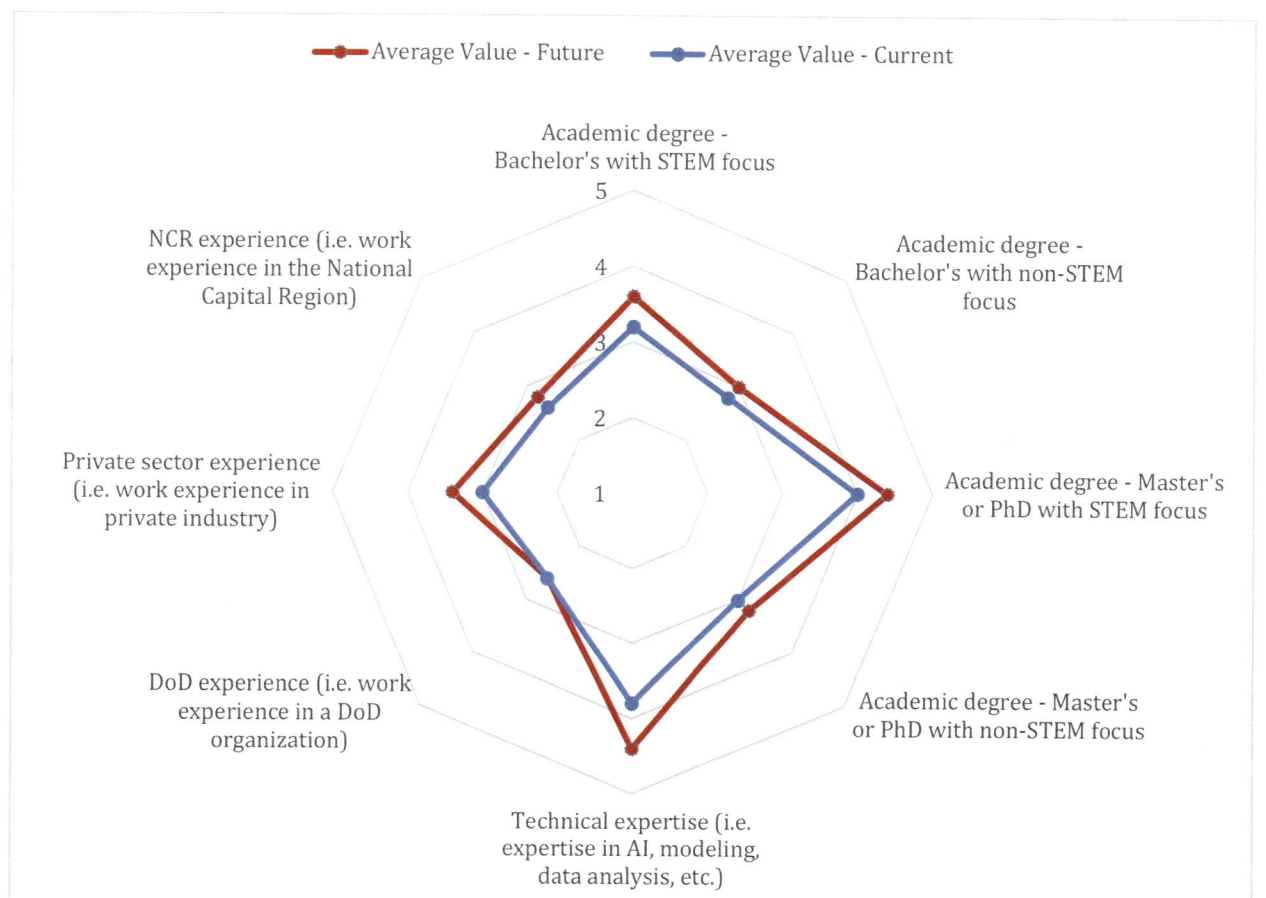

Contrary to some Air Force Futures interview findings, survey participants valued NCR and DoD experience lowest among the characteristics assessed. This reflects the possibility that while participants value those qualities as mechanisms to overcome hiring challenges, as discussed during the interviews, in an unconstrained environment presented through this survey they value other qualities more.

For example, according to one survey participant, who provided higher ratings for current and future technical expertise compared with current and future DoD experience,

> In general, the character of warfare continues to shift with society. Thus, some emerging technologies will impact the spectrum of strategy and we'll need a workforce that understands both the technical and non-technical implications. Ideally, a candidate will have a working knowledge of STEM issues as well as non-STEM issues with the ability to blend different levels of expertise for the A5/7. A greater need for understanding societal change and technology will become more important than heavy experience within the NCR.

Additionally, one survey participant who valued both current and future technical expertise remarked that "Air Force Futures is putting a strong emphasis on rapid integration of new and commercial technology, as well as model-based systems engineering to help with analysis of new strategies and concepts." This description matches interview findings that value civilian technical expertise in the technologist lane, which indicates that Air Force Futures desires civilians who can connect technological capabilities with strategic imperatives.

Overall, one survey participant emphasized Air Force Future's institutional desire for well-rounded civilians, saying,

> I would like to emphasize that there is no ideal candidate for the Air Force Futures. We thrive on having a diverse group, in every sense of that word. Some jobs require subject matter expertise (Combat Air Forces and weapons development come to mind), while others require expertise in narrative and communications (our strategic communications and Congressional relations shops). We require fresh thinking within a strategic framework, the ability to think critically, argue persuasively but professionally, and make decisions and come to agreement based on data and analysis, not passion. Those qualities are universal to Air Force Futures and would be ideal for all our candidates, adding technical prowess and subject matter expertise for specific jobs.

Conclusion

Overall, developing and fielding surveys to elicit demand and supply information from individuals at the work unit level with sufficient knowledge of the skills needed appears promising. Our pilot surveys have successfully identified key technical skills and provided valuable insights into respondent priorities and the personnel needed for mission success.

However, further development of a survey methodology that can be widely applied is necessary. It is crucial for individuals at the work unit level to possess a foundational understanding of workforce planning. This includes recognizing the skills essential for mission accomplishment, understanding the current workforce's capabilities, identifying methods for skill acquisition, and conceptualizing future workforce needs. Without a thorough grasp of these factors, assessments of workforce characteristics cannot effectively address supply and demand challenges. Facilitated discussions may be needed to interpret survey responses accurately.

Regardless of the survey methodology employed, it should be administered by individuals knowledgeable in job analysis, hiring processes, civilian personnel management, and data systems. Human Resources Specialists (occupational series 0201), which has been identified as a critical and hard-to-fill occupation, possess this expertise.[6] Fully resourcing and staffing these specialists as consultants and advisors could significantly enhance the ability of work units to identify their needs and secure the STEM talent necessary for mission success.

---

[6] Groeber et al., 2021.

Chapter 6

Strategies for Filling Gaps in STEM Competencies

The landscape for developing technical competencies within the DAF is diverse and multifaceted. Once civilians begin their careers in the DAF, they have access to a variety of professional development opportunities, ranging from on-the-job training to formal educational programs, internships, and specialized certification courses.[1] In this chapter, we begin by discussing options for strengthening the development of civilian technical talent. Next, we present different strategies that could be considered to address the training and education needs using the following four technical competencies identified as gaps in our interview and survey results: (1) digital modeling and engineering, (2) AI and ML, (3) data science, and (4) RF. These competencies are examples of the gaps that surfaced and were selected to reflect different levels of specificity (e.g., RF is more specific than AI) and the range of internal DAF and external resources available for training and education.

Enhancing Strategic Use of STEM Pipelines

The DAF offers extensive professional development opportunities in both STEM and non-STEM fields, fostering early interest through programs and supporting educational growth with tuition assistance and certifications. Institutions such as Air University and the Air Force Institute of Technology (AFIT) provide advanced education, while programs such as the Civilian Tuition Assistance Program and the Science, Mathematics, and Research for Transformation (SMART) Scholarship enhance further learning. However, our review indicates that many of these opportunities are optional. Most programs are guided by moderate central planning and are driven primarily by career fields, with promotion being determined by career field managers rather than being part of a broader strategic directive. In addition, some programs rely entirely on self-direction or initiation by individual units, as identified by SMEs through case study interviews (as discussed in Chapter 4 of this report). This suggests a potential for enhanced centralized planning and leadership engagement to better align education and training programs with organizational goals, particularly in addressing the gaps between STEM supply and demand for specific skills.

While this flexibility allows members to choose their career development paths, it may lead to underutilization of programs in fields where they are most needed. As DAF's STEM initiatives evolve, leveraging its extensive network of development opportunities and actively directing members to participate in programs aligned with anticipated needs could be beneficial. In STEM areas such as

---

[1] Appendix E contains a high-level survey of the types of education and development opportunities available and the segments of the workforce for whom they are targeted.

nuclear technologies, the USAF has implemented targeted professional development programs that span entire careers, and this model, as discussed in Appendix E, could serve as a blueprint for other technology areas.

Competency Focus: Digital Modeling and Engineering

Digital models are computer-readable representations of objects, phenomena, processes, or systems.[2] They can form virtual prototypes of weapon systems that have not yet been manufactured into "digital twins" that mirror and predict the activities and performance of a physical counterpart.[3] DE uses digital models to create and test systems in virtual environments, with the goal of reducing the need for physical prototypes and enhancing system monitoring and sustainment.[4] DE is closely related to model-based systems engineering (MBSE), which replaces documents with system design models, and modeling languages such as Systems Modeling Language (SysML), which help centralize and standardize information within a development team.[5]

DoD has stated that DE can enhance decisionmaking, reduce development time, and lower costs for weapon systems.[6] However, DoD lags behind the commercial sector and could soon lag behind U.S. adversaries in its adoption of DE.[7] Thus, DoD has initiated efforts to incorporate DE into its engineering practices.[8] For example, DoD has required its programs to incorporate DE[9] and established a Digital Engineering, Modeling, and Simulation Body of Knowledge to assist its personnel with implementing DE in their programs.[10] In addition, the military departments have begun incorporating DE into their engineering and acquisition practices and using DE to develop weapon and other systems.[11] For example, the Air Force has conducted a "digital fly-off" to compare candidate engines for the B-52 using computer simulations.[12]

The Systems Engineering Research Center (SERC), a DoD university-affiliated research center, developed a Digital Competency Framework (DECF) to identify DE competencies needed by

---

[2] DAU, "DAU Glossary of Defense Acquisition Acronyms and Terms," webpage, undated-c.

[3] Under Secretary for Research and Engineering, "Organizational Highlight: Digital Engineering, Modeling and Simulation," webpage, February 2024.

[4] Defense Business Board, *Creating a Digital Ecosystem*, Business Transformation Advisory Subcommittee, DBB FY24-03, February 29, 2024.

[5] Caitlyn Singam and Jeffrey Carter, "Model-Based Systems Engineering (MBSE)," Guide to the Systems Engineering Body of Knowledge, International Council on Systems Engineering, May 6, 2024.

[6] DoD, *Digital Engineering Strategy*, Office of the Deputy Assistant Secretary of Defense for Systems Engineering, June 2018.

[7] Defense Business Board, 2024.

[8] DoD, 2018.

[9] Department of Defense Instruction 5000.97, *Digital Engineering*, Office of the Under Secretary of Defense for Research and Engineering, December 21, 2023.

[10] Digital Engineering Body of Knowledge, homepage, undated.

[11] Jen Judson, "US Army Moves Out on Digital Engineering Strategy," *Defense News*, June 19, 2024; Department of the Navy, *United States Navy and Marine Corps Digital Systems Engineering Transformation Strategy*, 2020; J. Kyle Hurst, Steven A. Turek, Chadwick M. Steipp, and Duke Z. Richardson, *An Accelerated Future State*, Air Force Materiel Command, 2023.

[12] Stephanie Possehl and Philomena Zimmerman, "Digital Engineering and Modeling and Simulation," U.S. Department of Defense, Office of the Under Secretary of Defense, Research & Engineering, January 7, 2022.

DoD's engineering acquisition workforce.[13] Building on engineering competency models from nine organizations, SERC identified six foundational competencies needed to enter the DE workforce and 25 additional competencies categorized into five groups (Table 6.1). DECF includes a mix of technical competencies (e.g., Artificial Intelligence and Machine Learning, Software Engineering) and nontechnical competencies (e.g., Project and Program Management, Communications). For each competency, DECF describes five proficiency levels, from awareness to expert. This framework informed our assessment of DE resources.

Table 6.1. Systems Engineering Research Center Digital Engineering Competency Framework

Competency Group	Competency
Foundational	• Digital Literacy • Digital Engineering Value Proposition • DoD Policy/Guidance • Coaching and Mentoring • Decisionmaking • Software Literacy
Data Engineering	• Data Governance • Data Management
Modeling and Simulation	• Modeling • Simulation • AI/ML • Data Visualization • Data Analytics
Digital Systems Engineering	• Digital Architecting • Digital Requirements Modeling • Digital Validation and Verification • MBSE
Engineering Management	• Digital Model-Based Reviews • Project and Program Management • Organizational Development • Digital Engineering and Policy Guidance • Configuration Management
Systems Software	• Software Construction • Software Engineering
Digital Enterprise Environment	• Digital Environment Development • Management • Communications • Planning • Digital Environment Operations • Digital Environment Support

---

[13] Nicole Hutchinson, Kara Pepe, Mark Blackburn, Hoon Yan See Tao, Dinesh Verma, Cliff Whitcomb, Rabia Khan, Russell Peak, Adam Baker, "WRT-1006 Technical Report: Developing the Digital Engineering Competency Framework (DECF)—Phase 2," Stevens Institute of Technology, Systems Engineering Research Center, SERC-2021-TR-005, March 23, 2021.

Competency Group	Competency
	• Digital Environment Security
← Proficiency Levels: Awareness, Basic, Intermediate, Advanced, and Expert →	

SOURCE: Adapted from Hutchinson et al., 2021.

To help the DAF understand available resources for building DE competencies, we scanned DE resources from three sources: institutions of higher education (IHEs), massive open online course (MOOC) platforms, and DAU. We reviewed websites of IHEs in the SERC network and media describing DE or MBSE initiatives at these institutions. We also reviewed courses on DE, digital modeling, and MBSE from nine MOOC platforms, which were selected for their variety of technology-focused courses and large user bases or because they partner with Digital University, an Air Force initiative described later in this section.[14] We used the same procedure to review courses from DAU's catalog. Across the three sources, we focused on master's degree, certificate, and other programs that would be accessible for working professionals. Our scan was necessarily cursory, as systematic sampling and review of DE resources was beyond the scope of this project. However, it provides a useful picture of available DE resources for upskilling and reskilling.

Table 6.2 presents examples of DE-focused programs and courses offered by IHEs, MOOC platforms, and DAU. Many IHEs in our scan offered master's degrees or certificate programs in systems engineering for working professionals as well as isolated courses related to DE, such as introductions to MBSE or SysML. However, few IHEs offered programs that focus specifically on DE or related practices. We found several recent examples of IHEs building DE-focused programs using public funds.

For example, Purdue University used a National Science Foundation grant to develop a series of online MBSE modules,[15] and the University of Massachusetts Lowell partnered with Hanscom Air Force Base and the Massachusetts Development Finance Agency to create a custom DE program for Air Force engineers.[16] Auburn University has received several DoD grants to build DE and

---

[14] The platforms were Cloud Academy, Coursera, DataCamp, EdX, LinkedIn Learning, MIT OpenCourseWare, Pluralsight, Udacity, and Udemy. Digital University's landing page names Cloud Academy, Coursera, DataCamp, Udacity, and Udemy. We were unable to directly review courses offered on Digital University because the site can be accessed only by DoD personnel. The Digital University landing page also names Hack EDU, but we were unable to access course descriptions on that platform's website.

[15] "Development, Deployment, and Evaluation of Instructional Modules for Current and Future Practitioners of Model-Based Systems Engineering," webpage, undated.

[16] Jessica Casserly, "Partnership Delivers Specialized Training for Hanscom Engineers," 66th Air Base Group Public Affairs, February 28, 2022.

Table 6.2. Examples of Digital Engineering–Focused Programs or Courses

Provider	Program or Course	Description
Missouri University of Science and Technology	Graduate certificate in DE	Four-course certificate offered on campus or through distance learning; requires a college degree in engineering or basic science[a]
Naval Postgraduate School	Master's degree in systems engineering	Sixteen courses, including DE and MBSE; on-site and online instruction; requires college degree in engineering or related discipline; designed for Navy and DoD organizations[b]
Purdue University	MBSE Foundations and Applications to the Production Enterprise	Seven asynchronous instructor-facilitated online modules, including DE, MBSE, and SysML; can be taken together or separately depending on learner interests and responsibilities[c]
University of Massachusetts Lowell	Graduate certificate in DE	Four-course certificate developed for Hanscom Air Force Base;[d] provided to one cohort of 20 Air Force civilians, most with engineering background and title of systems engineer[e]
Coursera	Digital Manufacturing and Design Technology Specialization	Nine-course certificate, including MBSE and digital threads; content from State University of New York; covers digital advances in factories; for learners at all levels[f]
	Introduction to MBSE	Five-course certificate; content from Siemens; introduces MBSE software and principles; for advanced high school students, college students, and professionals[g]
DAU	Models, Simulations, and Digital Engineering	Five-hour online training; explains DoD policy on models, simulations, and DE and introduces MBSE and digital twins[h]
	Digital Acquisition Modeling Workshop	Self-paced workshop introducing MBSE by walking through development of a hypothetical weapons system; emphasizes SysML diagrams[i]
	Digital Twins for Predictive Maintenance Fundamentals	Online course, length to be determined; provides an understanding of how digital twins can be used to manage maintenance of a system[j]

[a] Missouri University of Science and Technology, "Systems Engineering," course catalog, undated.
[b] Naval Postgraduate School, "Systems Engineering," webpage, undated-b.
[c] Purdue University, The Robert H. Buckman College of Engineering Online Education Program, "Model-Based Systems Engineering Foundations and Applications to the Production Enterprise," webpage, undated.
[d] Casserly, 2022.
[e] Kavitha Chandra, Sara Kraemer, Emi Aoki, Flora Stecie Norceide, and Ola Batarseh, "Integrating Model-Based Systems Engineering and Systems Thinking Skills in Engineering Courses," paper presented at 2024 ASEE Annual Conference & Exposition, June 2024.
[f] Coursera, "Digital Manufacturing & Design Technology Specialization," webpage, undated-a.
[g] Coursera, "Introduction to Model-Based Systems Engineering," webpage, undated-b.
[h] DAU, "Models, Simulations, and Digital Engineering," course catalog, undated-f.
[i] DAU, "Digital Acquisition Modeling Workshop," course catalog, undated-d.
[j] DAU, "Digital Twins for Predictive Maintenance Fundamentals," course catalog, undated-e.

MBSE training programs[17] and hired MBSE-focused faculty, with the goal of growing its MBSE program.[18] These efforts, and the apparent lack of focus on DE at many IHEs, suggest that IHEs are in the early stages of integrating courses into coherent pathways for DE training and branding programs as DE.

Coursera and DAU offered several DE-focused programs or courses that appear to provide a basic or introductory level of knowledge. Coursera offered several four- to five-course certificate programs covering MBSE or other DE topics as well as stand-alone courses on software such as Autodesk Fusion and Siemens Solid Edge, which may be used for design, simulation, and development of products. Other MOOC platforms did not offer DE-focused courses with multiple modules and learning activities. These included six of the seven MOOC platforms named on the landing page of Digital University, a website launched in 2020 by the Air Force to offer technology-focused training from multiple MOOC platforms to USAF and USSF military and civilian personnel.[19] Other than Coursera, the six platforms appeared to focus on data management and analysis, software and web development, business skills, and cybersecurity, but they did not appear to offer DE courses.[20]

We identified approximately ten DE-related courses in DAU's course catalog, ranging in length and depth from a five-hour course covering DoD policy on to an 11-day workshop introducing MBSE by walking learners through the development of a hypothetical weapon system.[21] The course that appeared most technical—a digital twins for predictive maintenance course covering analysis and interpretation of sensor data to predict failures in a system—was described as offering a "fundamental understanding" of digital twin concepts and capabilities.[22]

Overall, our scan identified few DE-focused programs or courses among the training providers that we believed would be most likely to offer such resources. The programs we found included only one or two courses, such as a course on MBSE or SysML. These courses appeared to provide an introductory and relatively nontechnical level of knowledge and did not cover the full spectrum of competencies from SERC's Digital Engineering Competency Framework.

SERC's analysis of three DE resources is consistent with this assessment. They evaluated the competencies and proficiency levels provided by three DE resources: an MBSE program from Coursera; DAU's Models, Simulations, and Digital Engineering course; and MIT's four-course Architecture and Systems Engineering certificate program. They found that the Coursera and

[17] Cassie Montgomery, "DOD Invests $9.9M in Launch of Systems Engineering Technology Program," Auburn University Samuel Ginn College of Engineering, September 17, 2020; Dustin Duncan, "ICAMS Receives $9.2M to Further Model-Based Systems Engineering," Auburn University Samuel Ginn College of Engineering, March 13, 2024.

[18] Carla Nelson, "Recent ISE Doctoral Graduate Accepts Faculty Position," Auburn University Samuel Ginn College of Engineering, September 7, 2023.

[19] Laura Hayden, "Building Digital Literacy through Digital University," Air Combat Command Public Affairs, November 20, 2020.

[20] One platform, Hack EDU, appeared to be focused on cybersecurity, but we were unable to access course descriptions on its website. Workera, another partner named on Digital University's landing page, appears to provide learning assessments and monitoring rather than training.

[21] DAU, Course Catalog, webpage, undated-a.

[22] DAU, undated-e.

DAU resources would help learners attain "awareness" or "basic" proficiency in DE, but that seasoned engineers "would require substantially deeper training to become practitioners of digital engineering." They also found that the MIT program would provide "a solid fundamental understanding" of specific DE-related concepts and enable individuals to attain "advanced" proficiency in some cases.[23]

The building blocks of more comprehensive DE training programs may exist across a variety of departments within IHEs (e.g., courses that build AI and ML competencies can be found in computer science departments) but may not yet be packaged into coherent pathways and branded as DE. Recent initiatives to build DE programs suggest that some IHEs are beginning to create these pathways in response to DoD and industry demand.

Competency Focus: Artificial Intelligence

DoD defines AI as "the ability of machines to perform tasks that normally require human intelligence—for example, recognizing patterns, learning from experience, drawing conclusions, making predictions, or taking action—whether digitally or as the smart software behind autonomous physical systems."[24] DoD's 2018 AI strategy emphasized the critical importance of AI to military operations and the need for innovation to maintain a strategic advantage over adversaries. Central to this innovation is a workforce adept at leveraging AI capabilities.

DoD's Responsible Artificial Intelligence (RAI) Strategy and Implementation plan highlights the need to "build, train, equip, and retain an RAI-ready workforce to ensure robust talent planning, recruitment, and capacity-building measures, including workforce education and training on RAI."[25] This tenet builds on the AI education strategy published by DoD in 2020, in response to the 2020 National Defense Authorization Act (Section 256), which mandated the development of an AI education strategy for service members.[26]

AI competencies are to be maintained across several lines of effort: initiating workforce planning to ensure a highly skilled AI workforce, developing and updating DoD curricula on AI, coordinating with educational institutions such as DAU on key elements of AI training, and identifying potential external opportunities for competency development, including partnerships with academia and industry.[27]

DoD's strategy for enhancing AI competencies includes training and education in several key areas, as outlined in the 2020 AI education strategy. These areas include understanding the basic principles of AI, learning how AI can be applied to various military operations, developing the necessary infrastructure and software to support AI capabilities, implementing AI solutions

---

[23] Hutchinson et al., 2021.

[24] DoD, *Summary of the 2018 Department of Defense Artificial Intelligence Strategy: Harnessing AI to Advance Our Security and Prosperity*, 2018, p. 5.

[25] DoD, *U.S. Department of Defense Responsible Artificial Intelligence Strategy and Implementation Pathway*, DoD Responsible AI Working Council, June 2022, p. 31.

[26] DoD Joint AI Center and DoD Chief Information Officer, *2020 Department of Defense Artificial Intelligence Education Strategy*, September 2020.

[27] DoD, 2022, pp. 32–33.

effectively, and ensuring the workforce is capable of enabling and sustaining AI technologies.[28] By focusing on these areas, DoD aims to ensure that its personnel are well equipped to innovate and fully utilize AI capabilities and thereby maintain a strategic advantage over adversaries. The full list of competency topics and specific competencies named in the DoD's AI education strategy are listed in Table 6.3.

Table 6.3. Department of Defense Artificial Intelligence Education Strategy Competency List

Competency Topic	Competencies
1. Foundational Concepts	**Understanding AI:** Conceptualizing probabilistic reasoning and core elements of AI stack (to include Natural Language Processing, Natural Language Generation, Natural Language Understanding, Computer Vision, Neural Networks, Deep Learning, Robotics, and Autonomous Operations)
	Applying AI: Interpreting AI output and recognizing potential use cases, as well as understanding the basic requirements for successful application
	Advanced AI concepts: Understanding advanced and state-of-the-art AI methods
2. AI Applications: Opportunities and Risks	**Identifying trends:** Recognizing emerging trends in AI, as well as opportunities for research
	Identifying risks: Recognizing data and network security and privacy risks that come with AI, as well as AI bias, complementary compliance, incident response policies, and unique challenges to DoD (e.g., doctrine, warfighter displacement/dependence on machines, explainable AI, and trust)
3. Data Management and Visualization	**Managing data:** Understanding how to collect, store, and monitor data
	Visualizing data: Knowing how to structure and display data, as well as use data to create a story
	Data preparation: Preparing structured or unstructured data so that it is usable and meaningful to models
4. Responsible AI	**Operating ethically and legally:** Understanding the ethical issues related to AI and adhering to all relevant regulations
5. Infrastructure, Coding, and Software Development	**Programming and scripting:** Knowing how to code in languages that support AI tool development and data analysis (e.g., Java, Python, SQL)
	Software engineering: Understanding how to build effective software in the most efficient manner, including knowledge of DevOps, full stack development, and integration of established algorithms and pre-trained models

---

[28] DoD Joint AI Center and DoD Chief Information Officer, 2020, p. 8. (A full table of the competency model can be found on page 8.)

56

Competency Topic	Competencies
	Operating in cloud: Understanding various cloud services, cloud-native architectures, orchestration tools
	Computing: Understanding basic computing concepts (e.g., fog computing), and being able to differentiate different forms of computing
	Testing AI: Using models and prediction methods for evaluating AI performance
	DevSecOps: Understanding the tools and infrastructure needed to automate development, testing, securing, and deploying AI/ML-enabled software into the DoD
	AI frameworks: Understanding of the common frameworks used to implement AI methods
6. Mathematics, Statistics, and Data Science	**Performing analysis:** Applying mathematical and statistical analysis, (e.g., customized models/algorithms, predictive analytics) to understand and engage AI at technical level
7. AI Delivery	**Managing product development:** Understanding AI project management, including product development & prototyping
	Overseeing AI delivery: Understanding management of an AI delivery team, the structure and operating model, and effective planning, as well as how to facilitate the implementation of these tools by end users
	Leading AI strategy: Knowing best practice for implementing AI on a large scale as well as AI's impact on strategy
8. AI Enablement	**User-centric design:** Integrating Design Thinking, human-centered design, UX/HCI into system development & deployment
	Legal/IP Rights: Understanding of data rights, property rights, and intellectual property

SOURCE: Reproduced from DoD Joint AI Center and DoD Chief Information Officer, 2020.
NOTE: DevOps = development operations; DevSecOps = development, security, and operations; UX/HCI = user experience/human-computer interaction.

The education strategy further breaks down each AI competency topic into curriculum topics separated into basic (e.g., Intro to AI/ML Concepts), intermediate (e.g., Natural Language Processing), and advanced (e.g., Learning Algorithms and Training Models) levels of proficiency, providing a general roadmap for the type of technical skills the department believes its workforce will need.[29] The specific curriculum topics at varying levels of learning depth are shown in Table 6.4.

DoD has emphasized the critical need to develop AI competencies within its workforce through various channels. The DAF has aligned with this directive by issuing its own annex to the broader DoD AI strategy, which focuses on five core areas. One of these areas explicitly addresses personnel management, highlighting the necessity to "recruit, develop, upskill, and cultivate" DAF's AI

---

[29] DoD Joint AI Center and DoD Chief Information Officer, 2020, p. 10. (A full table of the proficiency levels of various competencies can be found on page 10.)

workforce.[30] Additionally, the DAF strategy underscores the importance of partnering "with Joint, industry, and academic partners to foster cross-collaboration for training and tradecraft," thereby preparing its workforce to optimally utilize AI capabilities.[31] Interviews with various stakeholders have consistently identified AI competencies as a significant need within the civilian workforce. This feedback, coupled with the explicit focus of DoD and the DAF on AI as a critical component of future operations, underscores the importance of professional development in this area.

AI competency development can be achieved through IHEs, MOOCs, and DAU. Opportunities for AI education and credentialing are proliferating within higher education, spanning different levels of expertise and time commitments. Flagship research universities such as Carnegie Mellon and Stanford offer a range of AI educational programs. Beyond traditional B.S., M.S., or Ph.D. programs, Carnegie Mellon's School of Computer Science operates "custom courses" designed for work teams, ten-week online courses on various AI topics, and certificate programs as part of its Executive and Professional Education portfolio. Similarly, Stanford offers custom courses, online programs on topics such as generative AI, and in-person cohorts focusing on strategic decisionmaking and AI applications.[32]

The DAF has also established partnerships with academic institutions to better address its research and personnel development needs. Notably, the 2022 launch of the Department of the Air Force–Massachusetts Institute of Technology (DAF-MIT) AI Accelerator aims to tackle "the challenge of educating, cultivating, and growing a world-class AI workforce" for the department.[33] Recommended resources from the Accelerator include offerings from both MIT and Stanford, which cover foundational AI concepts and professional certificates.[34] Additionally, DoD's IHEs, such as the Naval Postgraduate School and AFIT, offer AI-focused certificates, short courses, and full degree programs, including an AI track in Electrical and Computer Engineering.[35] Within DAU, a certificate in AI foundations for DoD comprises six units that introduce and instruct fundamentals of AI.[36]

Nontraditional training providers also offer a variety of AI training programs. For example, a search for "artificial intelligence" on Coursera returns more than 500 courses, covering topics such as AI infrastructure and operations fundamentals, AI for software development, and digital transformation using AI.

[30] DAF, *The United States Air Force Artificial Intelligence Annex to the Department of Defense Artificial Intelligence Strategy*, 2019, p. 5.

[31] DAF, 2019, p. 6.

[32] Carnegie Mellon University, School of Computer Science, Executive & Professional Education, "Artificial Intelligence," website, undated; Stanford University, Human-Centered Artificial Intelligence, "Professional Education," website, undated.

[33] DAF AI Accelerator Public Affairs, "DAF-MIT AI Accelerator Tackles Challenge of Cultivating, Growing World-class AI Workforce," October 31, 2022. To read more, see DAF AI Accelerator, home page, undated-a.

[34] DAF AI Accelerator, "Education," website, undated-b.

[35] Naval Postgraduate School, "Academic Catalog," website, undated-a; AFIT, Graduate School of Engineering and Management, *Academic Catalog 2023–2024*, undated.

[36] DAU, "CENG 003 AI Foundations for the DoD Credential," website, August 12, 2024.

Table 6.4. Department of Defense Artificial Intelligence Competency Curriculum Topics at Three Levels of Depth

Competency Topic	Competencies	Basic	Curriculum Depth Intermediate	Advanced
1. Foundational Concepts	Understand AI Apply AI Advanced AI topics	• Intro to AI/ML concepts • Key Terms • Neural Networks & Deep Learning • Supervised/Unsupervised Learning • Autonomy • Current AI Uses in DoD • Data in AI	• Computer vision • AI Robotics • Natural Language Processing • Speech Recognition	• Military applications of AI • Learning algorithms and training models
2. AI Applications: Opportunities and Risks	ID trends, ID Risks	• Future AI Uses in DOD • Cyber Risks and Vulnerabilities • Bias in AI	• Technical Future AI Uses in DoD • Identifying Cyber Risks and Vulnerabilities	• Doctrine • Explainable AI • Trust
3. Data Management and Visualization	Manage, prep, and visualize data	• Data-Driven Decisions and Culture	• Visualization Tools • Data Preparation for ML	• Data Engineering and Orchestration • Data Warehousing
4. Responsible AI	Operate ethically	• Responsible AI Use throughout DOD (Intro)	• Responsible AI Use throughout DOD	• Technical Issues in Responsible AI (e.g., measuring bias and fairness)
5. Infrastructure, Coding, and Software Development	Program, SW eng., cloud, computing, testing, DevSecOps, AI frameworks	• Intro to Programming and Languages • Intro to DoD DevSecOps	• Programming and Languages	• Software Development • Cloud Engineering • Distributed Computing • AI Infrastructure • AI Computing Platforms

Competency Topic	Competencies	Curriculum Depth		
		Basic	Intermediate	Advanced
6. Mathematics, Statistics, and Data Science	Performing analysis	• Data Analysis • Elements of Data Science • Intro to Algebra and Calculus • Statistics & Probability	• Algebra and Calculus • Statistics and Probability • Data Analysis	• Analytic and Empirical Methods • Algebra, Linear Algebra, Calculus • Predictive Analysis • Principal Component Analysis • Machine Learning Theory
7. AI Delivery	Manage prod. dev, delivery, strategy	• DevOps • Agile and Innovative Leadership • Analytical Thinking	• Product Management • Structure of AI Delivery • Military Strategy with AI Tech	
8. AI Enablement	User centric design, legal/IP rights	• Design Thinking • Data Rights, Property Rights, and Intellectual Property	• UI/UX Design	

SOURCE: Reproduced from DoD Joint AI Center and DoD Chief Information Officer, 2020.
NOTE: ID = identify; SW = software; UI/UX = user interface/user experience.

60

In summary, the DAF has access to a wide array of resources for AI training and development across various levels of instruction and types of institutions, including traditional universities, online platforms, and DAU. These resources are crucial for equipping DAF personnel with the necessary AI competencies.

Competency Focus: Data Science

While not as extensively highlighted as AI at the strategic level within DoD, data science is integral to departmental guidance on professional development and talent management. DoD's 2020 data strategy emphasizes the need to "carefully cultivate our data talent" to maintain essential capabilities and highlights tasks such as "provide data skill training" and "establish centers for data engineering excellence."[37] The 2023 digital strategy further incorporates data as a crucial element alongside analytics and AI, advocates for the expansion of digital talent management, and emphasizes training and upskilling across data domains. DoD aims to "focus on upskilling and reskilling Service members and civilians," preparing them for roles like data architect, data steward, and user experience designer."[38]

In alignment with the broader DoD strategy, the DAF recognizes data science as a critical competency. This is evidenced by the establishment of a Chief Data Office in 2017, which integrated AI. DAF leaders stress the importance of leveraging data for strategic advantage, which necessitates enhanced talent management.[39] Strategic guidance, DAF priorities, stakeholder discussions, and survey responses consistently identify data science as an essential skill to be developed within the workforce. Opportunities for acquiring and updating data science skills are abundant for DAF civilian personnel.

Numerous leading universities offer courses in data science. For instance, the University of California Berkeley provides a Data Science for Leaders Program, while Columbia University's Data Science Institute offers certificates and degrees in data science; MIT offers a 12-week virtual course in applied data science.[40] Within DoD, AFIT offers master's and doctoral degrees in data science, requiring coursework in areas such as algorithms, applied statistics, and ML.[41] AFIT's Data Analytics certificate program, launched in 2020, underscores its commitment to developing data competencies.[42] The Naval Postgraduate School also provides certificates and courses in data science, including analytics and big data management.[43]

---

[37] DoD, *DoD Data Strategy*, September 30, 2020, p. 6.

[38] DoD, *Department of Defense Data, Analytics, and Artificial Intelligence Adoption Strategy: Accelerating Decision Advantage*, June 27, 2023, p. 13.

[39] Kim Crider, "Air Force Data Strategy," Headquarters U.S. Air Force, briefing, undated.

[40] Berkeley ExecEd, "Data Science for Leaders Program," website, undated; Columbia University Data Science Institute, "Education," website, undated; MIT Professional Education, "Applied Data Science Program: Leveraging AI for Effective Decision-Making," website, undated.

[41] AFIT, undated, pp. 150–155.

[42] U.S. Air Force, "AFIT Launches Data Analysis Certificate," website, July 15, 2020.

[43] Naval Postgraduate School, undated-a.

As they do with AI, online platforms such as Coursera offer numerous data science courses, covering topics such as analytics, data science utilizing R and Python, and statistical inference. Additionally, DAU offers courses on data foundations, management, and analysis.[44] These diverse educational opportunities demonstrate the extensive options available for DAF civilian professional development and training in data science.

Competency Focus: Radar

While AI and data science are prominently featured in strategic documents for both DoD and the DAF, technical competencies in radar remain essential for Air Force operations. Stakeholder interviews and survey responses frequently highlighted the need for enhanced development of radar skills among civilian personnel. In 2022, the Under Secretary for Research and Engineering emphasized the importance of integrating sensing and cyber functions, including radar's role in electronic warfare and communications.[45] DoD's FY 2022–2026 management plan specifically mentions radar as a key tool for threat detection and response.[46] Air Force publications also regularly underscore radar capabilities in air surveillance and national defense.[47] As a specialized technical area, radar exemplifies a competency requiring targeted training and development from various resources.

Radar engineering education, particularly for nontechnical experts, is less prevalent than programs in AI or data science. However, IHEs such as Georgia Institute of Technology and MIT's Lincoln Laboratory offer professional courses for technical managers, military officers, and DoD civilians needing radar skills.[48]

Within DoD's technical institutions, the Naval Postgraduate School and AFIT provide core technical courses on radar. The Naval Postgraduate School covers topics such as radar fundamentals, airborne radar systems, and radar cross-section prediction.[49] AFIT's Department of Electrical and Computer Engineering recognizes radar's "critical importance to the Air Force, Space Force, and the Department of Defense" and includes it as a key focus area.[50] Many of its courses, including Aircraft

---

[44] DAU, "Courses and Schedules," website, undated-b.

[45] Heidi Shyu, "USD(R&E) Technology Vision for an Era of Competition," Under Secretary of Defense for Research and Engineering, memorandum, February 1, 2022.

[46] DoD, *DoD Strategic Management Plan: Fiscal Years 2022–2026*, July 2022, p. 71.

[47] For example, see May 1, 2020, explanation of JSTARS radar system jointly operated by USAF and U.S. Army (Airforce Technology, "JSTARS— Joint Surveillance and Target Attack Radar System," webpage, May 1, 2020) or USAF September 23, 2021, announcement regarding optimization of LRR (Deb Henley, "84th RADEES Optimizes Nation's LRR Systems for Air Surveillance, National Defense," 505th Command and Control Wing Public Affairs, September 23, 2021).

[48] For example, see Georgia Tech's Radar Systems Engineering course in its professional education program (Georgia Tech, Professional Education, "Radar Systems Engineering," webpage, undated) and MIT's introductory and graduate level radar systems courses (Massachusetts Institute of Technology Lincoln Laboratory, "Radar: Introduction to Radar Systems—Online Course," webpage, undated-b; Massachusetts Institute of Technology Lincoln Laboratory, "Radar: Graduate Level—Online Course," webpage, undated-a.)

[49] Naval Postgraduate School, undated-a.

[50] AFIT, undated, p. 76.

Combat Survivability, Fundamentals of Radio Frequency Analysis, and Advanced Topics in Radar Applications, focus on technical radar competencies.[51]

Due to its specific technical nature, radar-related courses are scarce on open learning platforms like Coursera, and DAU does not currently list radar-focused offerings. Given radar's specialized military applications, training opportunities are not as widely available as they are for disciplines with broader societal applications. The Air Force may face constraints in training options for radar skills, necessitating "in-house" development, while it can leverage more traditional educational opportunities for other competencies through universities and online platforms.

Conclusion

Our analysis revealed numerous training options for three of the four competencies identified as having gaps by interview participants: DE, AI, and data science. These options are available at IHEs, on MOOC platforms, and through DAU.

However, SERC's report suggests that many DE programs cover only a fraction of DE competencies and would not help a learner advance beyond basic of proficiency. Descriptions we found of efforts to build DE programs suggest that the different kinds of courses needed for DE competency have not yet been integrated into comprehensive and in-depth programs that are branded as DE. In the terms of SERC's DE competency framework, such training options appear to be geared toward instilling an "awareness" or a "basic" level of proficiency in STEM competencies.

In contrast, training options for RF are limited. This disparity likely exists because DE, AI, and data science have broad military and commercial applications, while radar is a more specialized field with specific military uses.

Many training options appeared to be accessible for working professionals. Examples include master's degrees and certificates offered by IHEs, including military IHEs such as AFIT, and individual courses or short programs offered by IHEs or private companies (e.g., Siemens) through MOOC platforms. However, some training options appeared to be relatively nontechnical and focused on management or policy-related aspects of weapons or other technologies.

The developmental opportunities discussed, such as those in DE, AI, data science, and RF, are indeed available for Air Force civil servants. They can access these programs through partnerships with educational institutions, online platforms such as Coursera, and internal resources such as DAU and AFIT.

To ensure that civil servants have access, the Air Force could enhance awareness by promoting these programs through internal communication channels, offering guidance from supervisors and development teams, and integrating these opportunities into career development plans. In addition, establishing partnerships with universities and expanding in-house training programs could further support skill development.

---

[51] AFIT, undated, pp. 205–223.

Chapter 7

Key Findings and Recommendations

The objective of this project was to investigate the DAF's need for STEM talent within its civilian workforce, which we explored primarily through three case studies. We found it very difficult to identify clear sources of demand and supply for civilian STEM talent—particularly sources of information that would aggregate across the DAF. Although various data systems provide fragmented data, they do not collectively offer an accurate picture of technical requirements for civilian personnel or a clear account of current workforce talent.

Our key finding is that the most accurate and accessible information on STEM workforce demand and supply resides at the local level where the nuance of different missions can be more readily translated into specific requirements and where there is flexibility to quickly address critical needs when gaps in technical talent arise. While this local focus may create a disconnect with strategic workforce functions and the desire for centralized talent tracking systems across the DAF, our recommendations prioritize supporting local organizational levels and then, over time, evolving into a comprehensive, integrated talent management system.

In this chapter, we summarize the project's key findings and outline recommended actions for the DAF.

Key Findings

What Do We Know About the Demand for Technical Talent?

- Existing DAF data systems and personnel practices (e.g., job analyses, PDs, competency frameworks) are incomplete and may provide misleading signals of demand for technical talent.

  - Information from these sources does not necessarily reflect the actual tasks employees perform in their positions.
  - Position announcements rarely specify skills and are not tailored to individual positions. Instead, they emphasize general requirements for an occupational series.
  - Specific, technical skills related to emerging technologies, such as AI, and senior leadership expectations were not present in DAF job announcements.
  - Tailoring job announcements and PDs is labor-intensive and can cause processing delays. As a result, job announcements and PDs are becoming more generalized and standardized, resembling a standard core personnel document (SCPD).[1]

---

[1] An SCPD is an "off the shelf" PD that is considered accurately classified and stored in the Air Force Personnel Center's SCPD Library. SCPDs are intended to "eliminate duplication of effort in composing individual descriptions and eliminate confusion arising from variations in phraseology that do not represent variations in substance." DAFI 36-1401, *Civilian Position Classification*, Secretary of the Air Force, May 22, 2023.

- DAF organizations have distinct mission sets, organizational histories, and sizes, all of which shape civilian billet structures and STEM talent needs.

  - Some organizations may require more technology awareness and less deep technical expertise.
  - Deep technical expertise may be needed by some organizations temporarily to solve specific "sticky" problems.
  - SMEs indicated a demand for technical experts who also possess operational expertise.

- The demand signal for technical competencies is localized and not well aligned with strategic workforce functions.

  - Supervisors address needs as they arise, often influenced by gaps in funding for authorized positions.
  - The number of personnel required depends on the mix of competencies and proficiencies in the workforce. Fewer personnel may be needed if they possess high proficiency or a broad set of skills compared with a larger number of specialized personnel with lower proficiency.

- Case study SMEs identified several competency demands, some spanning organizations and others unique to a particular organization or mission.

  - Broadly relevant competencies included AI, ML, and data science; various engineering disciplines including electrical engineering and systems engineering; and cyber.
  - More specific competencies included digital modeling and RF.

What Do We Know About the Supply of Technical Talent?

- Knowledge of the supply of civilian personnel with specific skills is localized with few methods for tracking them.

  - Aggregate statistics on civilian STEM talent in the DAF do not provide insights into gaps in the supply.
  - The occupational series held by an individual is an insufficiently detailed designator for tracking specific skills.
  - Academic degrees are one indicator of the technical skills an individual possess, but they are not sufficient.
  - Competency systems or frameworks which can be useful for tracking skills are either not used, not accurate, or used inconsistently.

- The level of expertise within a particular skill set is often self-reported or not reported at all.

  - We found no recognized system by which levels of expertise are broadly measured or understood.
  - We did hear from SMEs that technical skills exist on a spectrum, ranging from basic technological awareness and tinkering to the application of technology and deep technical expertise, and because they do, accurate reporting is challenging.

What Do We Know About the Gaps in Technical Talent?

- Determining gaps in technical talent is complex and how to do so remains unclear.

  - Civilian positions can be authorized but not funded or funded but not filled. These situations do not necessarily indicate a gap or increased organizational risk, as supervisors can find ways to fill critical needs.
  - Gaps can manifest as insufficient numbers of personnel in an occupational series, lacking specific competencies, or having inadequate proficiency levels.

- Gaps arise from various causes, including personnel being moved from one program to another, unfunded positions, delays and barriers due to the civilian hiring system, and short-term needs for specific expertise.

How Does the Department of the Air Force Currently Address Gaps in Technical Talent?

- Gaps in STEM talent are addressed using multiple talent pools.

  - When permanent civilians cannot be hired, organizations use program funding to fill open positions with qualified contractors.
  - Depending on the specific needs, hiring managers can also coordinate with other DAF organizations to facilitate internal temporary or permanent moves. Internal transitions may also occur when DAF priorities shift and civilians need to be placed following program closures.
  - In some cases where an organization faces a particularly difficult but temporary problem requiring deep domain expertise, they may leverage talent from FFRDCs.

- Organizations focus on hiring recent college graduates with the right soft skills and then invest in their training and development.

  - SMEs indicated that new employees could learn the required technical skills through on-the-job training and opportunities to gain relevant experience.
  - This strategy is deemed necessary because of the small size of the applicant pool with desired technical skills and experience, which is attributed in part to higher salaries offered by private industry.

- When available, the DAF addresses gaps in technical talent through diverse functional development opportunities.

  - The DAF emphasizes cultivating a broad range of technical knowledge and skills within the civilian workforce, prioritizing the hiring of motivated individuals who can be trained over time to meet specific needs through on-the-job training and tailored educational initiatives.
  - Opportunities for developing skills include on-the-job training, formal education, internships, and specialized certification courses.

- The range of training and education opportunities vary across skills, with more options available for AI and fewer options for specific technical skills such as radar.
- In some technical specialties, the DAF leverages partnerships with educational institutions such as AFIT and DAU.

What Cross-Cutting Issues Affect Workforce Management?

- Information about the civilian workforce—the demands for and the supply of technical talent—is largely decentralized to local, individual organizations and work units.

  - As a result, determining personnel supply and workforce demands cannot be effectively summarized across different occupational series, personnel systems, or MAJCOMs.
  - Current DAF personnel systems are not designed to aggregate this localized information from organizations and work units.

- The term *STEM* has multiple definitions and may not be the right unit of analysis for estimating the current and future workforce.

  - Definitions of *STEM* are generally limited to occupational series or academic degree and provide limited information about the required or possessed technical competencies or skills.
  - Systematically focusing on STEM competencies could offer more detailed information to support workforce planning (e.g., necessary training) but may be challenging for supervisors (e.g., branch chiefs) who are unfamiliar with or do not use competency frameworks.

- The civilian personnel system creates barriers to hiring talent, which are not limited to STEM occupations.

  - Salaries are often not competitive with industry positions.
  - Changing an occupational series on a PD requires reclassification, which adds significant time (e.g., six months or more) to the review process and hiring.
  - Requests to post job announcements are frequently returned for what appear to be minor discrepancies or errors, causing civilian personnel functions to be perceived as gatekeepers rather than supporting the work units who are hiring.
  - If requirements in a job announcement are too specific, finding qualified candidates may be too difficult; therefore, hiring managers must balance the need for specific skills with the flexibility required to ensure that critical vacancies are filled.
  - Civilian personnel processes create real and perceived barriers to hiring STEM talent into the appropriate occupational series. For example, SMEs indicated that instead of specifying a need for electrical engineers (0850 series), they post job announcements for general engineers (0801 series) to provide the most flexibility for filling a position.

- LLMs (e.g., GPT 4.0) and ML are useful tools for augmenting human analysis of text documents (e.g., PDs) but are not yet sophisticated enough to replace SMEs. Specifically,

LLMs can be used to support personnel supply and demand analyses by extracting text, identifying technical skills, and recommending terms to include in traditional NLP analyses.

- Monitoring skill demands is most effective when target skill labels, acronyms, and related labels are known and well-defined. LLMs and ML models can aid in the development and standardization of these labels.
- The quality of skill information extracted by LLMs depends on the accuracy and completeness of available employee and position data.
- Extracting information about employee proficiencies and corresponding required proficiency levels for specific skills requires SME input.

Recommendations

The study findings point to three core actions for the DAF. The recommendations begin by emphasizing support for local organizational levels, then address broader strategic considerations to close skill gaps and invest in integrated talent management. These recommendations are cross-cutting as they could have an impact on issues related to technical supply, demand, and gaps analysis.

Focus Resources and Support at the Local Level

Emphasize resources, training, and policies to support local workforce management. While centralized systems for tracking talent across the DAF are desirable, the civilian workforce is managed at the local level. Achieving accurate macro-level perspectives may be challenging due to existing constraints and distinct personnel systems. Therefore, support should be directed to local units to enhance their hiring, training, and workforce planning efforts. Additional assistance could help local levels document and communicate their workforce supply and demands to higher organizational levels.

Enhance support for local units through strategic engagement. To effectively support hiring managers and work units, functional and personnel representatives at higher organizational levels (e.g., AFLCMC's Personnel Directorate and Engineering Directorate for divisions in AFLCMC) must shift their perspective to view local units as their primary customers. This requires a change in emphasis from merely enforcing hiring rules to actively assisting local units in acquiring the right talent and training. Currently, these offices focus on procedural compliance, which often delays hiring processes over minor issues, without maintaining essential information on hard-to-fill skills or providing structured recruitment support. Instead, they should facilitate targeted recruitment and training by aggregating needs from local units and helping to identify quality ranking factors that communicate to the DAF and to potential applicants what technical skills are valued.[2]

Recognizing that local managers are primarily engineers, scientists, and program managers with mission-specific roles, these offices should assume responsibility for workforce management tasks,

---

[2] As stated by OPM, "Quality ranking factors are KSAs/competencies that could be expected to significantly enhance performance in a position, but, unlike selective factors, are not essential for satisfactory performance." See OPM, "General Schedule Qualification Policies: General Schedule Operating Manual," webpage, May 2022.

thereby allowing local units to concentrate on their core objectives. While some support is already provided, additional resources may be needed to enhance effectiveness. This recommendation aims to improve existing support and focus on acquiring the technical talent essential for mission success at the local level.

Communicate to supervisors the purpose of tracking technical skills. Clearly defining and communicating the purpose of tracking technical skills will guide hiring managers and supervisor efforts. Supervisors need to understand the importance of documenting current and future needs for specific occupational series, proficiency levels, and skills. The message may be particularly important when divisions have unfunded positions or believe requests for additional resources will not be supported.

Support local efforts to identify critical technical skills. Local units possess a clearer understanding of their specific needs and the risks associated with unaddressed gaps. However, they may require guidance and support in documenting current and desired technical skills. Human resource specialists can help to provide this support. It is essential to address current and anticipated shortages in human resource specialists and ensure they receive adequate professional training and development to facilitate more effective workforce planning.[3]

To identify critical skills within local work units, we recommend that human resource specialists review PDs and job analyses to develop an initial list of skills and definitions. Once an initial list has been compiled, the human resource specialist should meet with branch and division-level SMEs to review and update skills and definitions. The focus should be on critical skills that contribute to mission effectiveness and may benefit from further monitoring or development. As needed to support broader communication of skill gaps, human resource specialists can coordinate across work units and with career field teams to aggregate skill supply and demand information from the local work units they support.

Identify and Evaluate Mechanisms to Close Specific Technical Skill Gaps

Identify which technical skills need development and assess whether the necessary training is available. For certain technical skills, consider external training and education programs to build foundational knowledge and the required level of proficiency (e.g., awareness versus expert; see Chapter 6). Custom training may be needed for skills that are more specific to a particular organization or mission. Start by reviewing courses offered internally by the DAF. To gain a comprehensive perspective, it may be helpful for DAF science and technology stakeholders and SMEs (including the office of the AF/ST, chief scientists in other organizations, AFRL and AFIT) to assist in the identification of needs and the development of curricula to fill the gaps in support of functional leads. If internal courses are insufficient, explore external options such as those at IHEs. Evaluate how widespread the skill needs are across different organizations. If multiple organizations require the skill and no relevant training is available internally or externally, consider developing a new curriculum.

---

[3] Groeber et al., 2021.

Consider Investing in the Building Blocks of a Talent Management System

A talent management system will streamline data collection and enhance the strategic alignment of workforce capabilities with DAF mission objectives. By focusing initially on the foundational elements, the DAF should identify the key data points that need to be captured and tracked. This approach will help determine what information can be obtained from existing systems, such as the DCPDS, to provide the necessary data for informed decisionmaking.

Enhance data collection on the civilian workforce. Explore methods to generate more comprehensive and accessible data about the civilian workforce. Without accurate and up-to-date information on employees' education, skills, and specific job requirements, the return on investment in data analytics and AI/ML may be limited. Improved data collection will enable better decisionmaking and more effective workforce planning.

To enhance data collection on the civilian workforce, the Air Force should modernize its data infrastructure by transitioning to cloud-based platforms with user-friendly interfaces, ensuring easier access for all stakeholders. Determining what data are needed and establishing standardized data elements and a robust governance framework can improve consistency and clarity. Increasing data granularity through a comprehensive skills taxonomy and regular updates could provide detailed insights into employees' skills and experiences. Ways to address data accuracy and completeness include automated validation processes and secure self-service portals for employees to update their information. Leveraging analytics, such as AI and ML, could enable the identification of workforce patterns and suggest predictive modeling to anticipate needs. Finally, implementing data literacy programs and providing ongoing technical support will be needed to ensure personnel can effectively utilize new systems and tools, thereby enhancing overall workforce planning and decisionmaking.

Consider developing a broad competency framework for technical skills. Tracking broad technical competencies across the DAF can be beneficial but determining the right level of specificity is challenging. Competencies that are too broad lack the detail needed to guide workforce planning efforts and will have wide ranging interpretations depending on the background of the SMEs completing the survey. For example, AI as a stand-alone competency is likely too broad to be meaningful for most DAF organizations. DoD's competency list for AI (Chapter 6) illustrates that understanding AI risks and opportunities requires different skills than writing and evaluating AI software.

Therefore, the first step towards building a competency model that is useful is to determine the level of analysis that is required to support the intended purpose for the model. If the purpose is to support training across multiple units DAF-wide or to support the internal labor market (e.g., internal transfers), then competencies should be broad but have a common interpretation across relevant SMEs. Several competency models have already been developed for different functional communities within the DAF and across professional organizations. To support building such a framework, we recommend coordinating with the branch chief, occupational competencies in the Air Education and Training Command as the core SME to provide support and direction. This office is currently focused on competencies for occupations, but we suggest focusing on competencies to support the development of specific technologies, which may span multiple occupational series.

Conduct periodic workforce surveys to identify supply, demand, and technical skills gaps. Focus on technical competencies that are important to the DAF's current and future missions. While no single authoritative source of technical competencies exists, multiple competency models can serve as starting points (e.g., functional career field models, White House critical and emerging technologies).[4] Once an initial list has been created, gather feedback from supervisors (e.g., at the branch and division level) to refine and finalize the list. Building a comprehensive list should not be the immediate goal, as it can be amended over time. Once a reasonable set of competencies is identified, the DAF can incorporate them into data collection efforts to better monitor workforce supply and demand through surveys, interviews, and discussions.

Regular surveys, such as those provided in Appendix D, will help keep the data current and relevant. The specific questions can be tailored to meet defined purposes. We recommend that local work unit supervisors coordinate with their human resource specialists to support data collection about the technical skill requirements and supply in their workforce.

Finally, we emphasize the importance of piloting any new questions or combinations of questions to ensure survey respondents understand their purpose and the information being sought. Follow-on focus groups with work unit supervisors should be conducted to review results and ensure accurate interpretation of skill requirements and gaps.

In some cases, more specific information about the skills and proficiency levels held by each employee within a work unit for a specific career field may be needed. These survey efforts are resource-intensive but may provide insights about potential gaps between current and desired proficiency levels. To conduct these types of surveys, we recommend coordinating with the Defense Civilian Personnel Advisory Service to administer and analyze results from DCAT.

Adopt a holistic approach to workforce planning. Rather than focusing solely on civilian personnel, the DAF should evaluate STEM talent more broadly, including uniformed service members, contractors, and personnel from FFRDCs. Supervisors indicate that they already consider these diverse sources when identifying and addressing skill gaps, suggesting that a more integrated strategy could enhance the DAF's ability to identify where investments in civilian technical talent are needed.

Conclusion

This report highlights the critical need for the DAF to enhance its approach to managing technical talent within its civilian workforce. Our investigation reveals that the most effective information on workforce demand and supply resides at the local level, where mission-specific needs can be effectively addressed. However, this local focus can create a disconnect with strategic workforce functions. To bridge this gap, we recommend prioritizing support for local units while gradually developing the building blocks for a more comprehensive talent management system. The DAF should consider starting with a limited number of specific technical areas and potentially developing pilot programs to refine these strategies.

---

[4] Fast Track Action Subcommittee on Critical and Emerging Technologies, 2022. For an example of competencies for the Engineering and Technical Management functional area, see Thomas Simms, "Engineering and Technical Management (ETM) Functional Area Framework Brief," Acting Director, Engineering Policy & Systems, Office of the Under Secretary of Defense for Research and Engineering, March 8, 2022.

Appendix A

Large Language Model Skill and Competency Extraction

We used the following methodology to extract and standardize competencies with an LLM.

- **Step 1.** We used the same groups of sample data from the NLx dataset as in the keyword analysis in Chapter 2, focusing on the DAF and defense-related engineering positions.
- **Step 2.** For each job announcement in each group, we used OpenAI's GPT-4 to preprocess the text into a string containing organization name, job title, and key phrases that contain competencies and knowledge areas relevant to the role based on the text of the job announcement. This extraction step was critical for removing irrelevant text (e.g., application instructions) and retaining text that describes KSAs, and related competency requirements. We used an automated script that called RAND's OpenAI API endpoint for each job announcement using consistent parameters and prompt language. The temperature variable, representing stochasticity in the output, was set to 0, making our output deterministic. The prompt used in this step is included in the first row of Table A.1.
- **Step 3.** Using the output from Step 2, we applied an extraction step to each job announcement to identify keywords and phrases associated with competencies. These were organized into three categories: management/supervisory, soft skills, and technical. The prompt used in this step is included in the second row of Table A.1.
- **Step 4.** We collected all terms in the technical category from Step 3 and ran a standardization script to allow comparisons among job announcements. The standardization process included the following steps:

    (a) After removing duplicates from the compiled list of technical terms, we used the OpenAI Ada embedding model to convert each term into an embedding vector, a high-dimensional sequence of numbers representing the syntax and content of a text string.

    (b) We calculated the embedding distance between two vectors w and v using the formula $1 - w^T v$ and then used this metric for a clustering algorithm with parameters chosen based on inspection of the clustering output.

    (c) Each cluster was assigned a label representing terms within it. This was done by selecting the most frequently occurring term as the cluster label, accounting for punctuation and case, and in the case of a tie, using GPT-4 to create a cluster label based on the cluster.

    The prompt used in this step is included in the last row of Table A.1.

Table A.1. Prompts Used in Large Language Models–Augmented Job Announcement Analysis

Step	Prompt
Preprocessing	"""Please read the position description carefully. First, list the organization name. Next, extract the text sections from the position description that are most relevant to identify position competencies, and then list them. Text sections should be relatively short. If there are no relevant text sections, write No relevant text instead. Do not provide any additional headers and definition of the technology. Extract the full sentence. If no relevant information is included in the PD, then write N/A: EXAMPLE OUTPUT. Please follow this format exactly. Do not add any unnecessary characters that would confuse Excel: | Air Force | | Troubleshoot and accomplish minor repairs, remove and replace components, and document all repairs in technician workbooks | Knowledge of engineering disciplines | Perform software, firmware, and hardware installation or upgrades on the system | | Army | | Knowledge of military installations | Test equipment to refine measurements | Position Description:\n """
Competency extraction	"""This is a position with extracted KSAO, task, and duty entries. Please read all of them carefully. Provide concise but descriptive labels for KSAOs. Use best practices to provide a label. Use existing operations research, O*NET, or other professional taxonomies to help determine the most specific and appropriate labels. Do not provide any additional headers or definition of the KSAO labels. Step 2: Review the KSAOs and organize into 3 headings. 1) Management/Supervisory, 2) Soft Skills, or 3) Technical. Separate each KSAO category and label with a | EXAMPLE OUTPUT. Please follow this format exactly. Do not add any unnecessary characters that would confuse Excel. Please present the KSAOs by numerical category starting with Management/Supervisory KSAOs and finishing with Technical KSAOs: 1) Management/Supervisory | Team Leadership | 2) Soft Skills | Communication | Collaboration | 3) Technical | Machine Learning | Neural Networks | SQL | Weapons Systems | Data Mining | Big Data Position extract: \n """ """Return a label which describes most or all the words in the following list, with a preference for shorter and more general labels. Your response cannot be in the form of a sentence. The list is: """
Standardization	"""Return a label which describes most or all the words in the following list, with a preference for shorter and more general labels. Your response cannot be in the form of a sentence. The list is: """

NOTE: KSAOs = knowledge, skills, abilities, and other characteristics; SQL = Structured Query Language.

Validation

There are several known limitations of LLMs and generative AI, including the potential for hallucination and poor reproducibility. Although the results could vary if the analyses were repeated, we took the following steps to evaluate the accuracy and quality of ChatGPT's output.

First, we assessed whether any of the extracted text was hallucinated by ChatGPT. We manually compared ten PDs with the extracted text and then executed a script across all announcements to identify any text in the extraction that did not occur in the original position announcement. Both the manual comparison and the automated search indicated that ChatGPT did not generate new content and only extracted text from the original position announcements.

Next, we evaluated the relevance of the competency labels generated by ChatGPT using two strategies:

- An internal SME (i.e., an engineer with a Ph.D.) evaluated the relevance of 100 competency labels generated by ChatGPT. The SME found that 77 percent of the labels were relevant.
- A second internal SME (i.e., an engineer with a Ph.D.) reviewed the full extracted text for 11 position announcements and all corresponding competency labels generated by ChatGPT. The SME evaluated the relevance of 107 competency labels and identified any critical competencies that may have been missed by ChatGPT. The SME identified two competency labels that were too vague to be meaningful. Another concern was that ChatGPT identified multiple competencies for which only one might be necessary (e.g., listing multiple engineering disciplines). Therefore, the competency labels generated by ChatGPT should be viewed as relevant but not necessarily all required. The SME also noted that one or two competency labels may be missing from the sets for three position announcements. These omissions were specific to knowledge of particular technologies (e.g., weapon systems, crude oil distillation) or specific types of engineering (e.g., physical security systems engineering). In future analyses, the ChatGPT prompt might be modified to explicitly capture knowledge areas for more specific technologies described in job announcements.

Case Study Interview Methodology and Protocol

The RAND team interviewed a total of 27 SMEs working in PEO Digital, HQ PACAF, and Air Force Futures during 27 one-on-one interview sessions. We conducted interviews with key Air Force representatives who provided insights and expertise on civilian personnel requirements, utilization, management, hiring, and STEM needs within their organizations. Interviewees were supervisors working at different levels within their respective organizations and represented perspectives from different organizational levels.

To identify interviewees, we received contact information from our project sponsor's office and from interviewees themselves when they suggested other personnel whom we should contact for interviews. We sent out email invitations for voluntary participation in our project interviews, providing potential interviewees general background information on the project and requesting they provide times to schedule a virtual, unclassified interview based on their availability.

We conducted all interviews over Microsoft Teams between December of 2023 and June of 2024. Each interview consisted of one Air Force participant and, from the RAND team, one interviewer, a note taker, and in some cases one to two additional project team members. The interviews were semi-structured and utilized a predeveloped interview protocol to guide the questions and discussion, and the RAND team provided both the interview protocol and informed consent documentation to all interviewees in advance of the scheduled discussion. When applicable, the RAND team asked interviewees to send any supporting documents to the RAND team after the interview for use as background material in support of this project.

The protocol used in the interview addressed general introductory questions on interviewee backgrounds; STEM-related requirements; STEM-related knowledge, education, and experience; potential skill gaps; future requirements; and closing questions requesting recommendations for other interviewees as well as any final topics that interviewees wanted to raise. Specific questions across these general categories are as follows:

STEM-Related Requirements:

1. Can you describe the general mission/responsibilities for managing technologies in your program/office?
2. What does your workforce mix look like (e.g., military, civilian, permanent, temporary)?
3. Can you describe the process that your program uses to identify personnel requirements?
4. What other data do you collect or use to support talent management in your program or office?

5. What data should be collected but may not be regularly collected that would help you estimate STEM-related demands in your program?

STEM-Related Knowledge, Education, and Experience:

6. Which functions in your program have the greatest requirements for STEM backgrounds? Can you give a brief overview of these functions?
7. All else being equal, would you prefer someone with an advanced degree and limited experience in your program/area or someone with extensive experience and a bachelor's degree?
8. How do you develop STEM competencies within your program or area?
9. For nontechnical positions in your program or area, please describe any useful or required STEM-related competencies.

Potential Skill Gaps:

10. How many positions are currently unfilled in your program or area?
11. Do you have any mission critical hard-to-fill positions in your program or area? If so, please describe.
12. How well does the quality of your personnel compare to available talent in the commercial sector?
13. Are there current gaps or imbalances in STEM knowledge or skills in your program or area?
14. What strategies, if any, has your program attempted to implement to close this gap (e.g., reassigning internal talent across programs or from other areas within the DAF)?

Future Requirements:

15. Are there any positions or competencies at risk in the future? If so, please describe.
16. If you were to receive additional authorizations, what occupations or competencies would you expand to improve mission capabilities?
17. If authorizations were to be reduced, which occupations or competencies would be critical to retain to avoid mission degradation or failure?

Appendix C

Supplemental Information on Case Studies and Case Study Populations

In this appendix, we present additional information about and analyses of our case study populations of PEO Digital, Air Force Futures, and HQ PACAF. This information supplements the background information and key insights presented in Chapter 4 of this report.

PEO Digital

We conducted 15 interviews with PEO Digital SMEs from programs within four specific divisions: Aerospace Management Systems (HBA); Theater Battle Control (HBD), Airborne Warning and Control System (HBS), and Aerospace Dominance Enabler (HBZ).

As of September 2023, there were 168 STEM civilian personnel in our focal PEO Digital divisions, comprising roughly 24 percent of all civilian personnel across those offices. Within PEO Digital, STEM personnel were concentrated in the four largest divisions by personnel count: Airborne Warning and Control Systems (26.9 percent of all STEM personnel), Aerospace Management Systems (26.8 percent of all STEM personnel), Theater Battle Control (23.6 percent of all STEM personnel), and Aerospace Dominance Enabler (22.7 percent of all STEM personnel). These four divisions comprised 88.6 percent of all PEO Digital personnel and 93.5 percent of all STEM personnel.

To contextualize STEM personnel as part of PEO Digital, Figure C.1 compares the percentage of individuals in PEO Digital who were STEM personnel with the percentage in other occupational groups. STEM personnel comprise roughly 24 percent of the PEO Digital civilian workforce, which is a smaller percentage than that of civilians working in business and industry occupations[1] and just above general administrative, clerical, and office services occupations.

Figure C.2 compares the percentage of STEM and non-STEM civilian personnel by GS-equivalent pay grade. A higher percentage of STEM personnel were in grades equivalent to GS-14 or GS-15, the highest grades on the civilian pay scale.

---

[1] The occupational group of business and industry includes occupational series such as contracting, purchasing, and financial analysis.

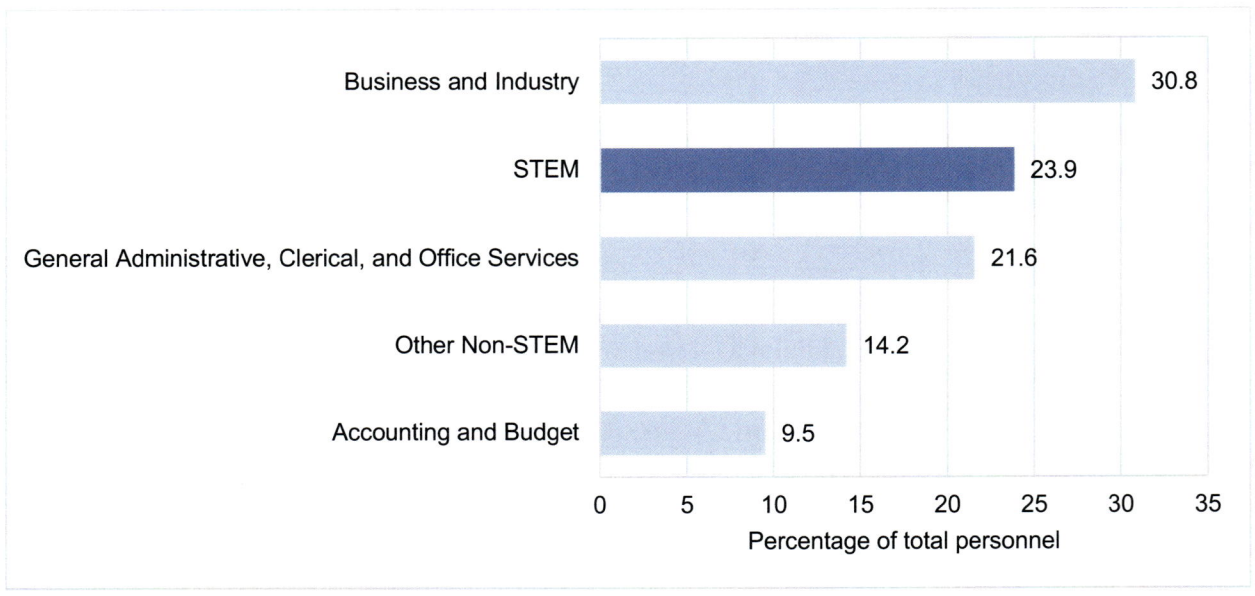

Figure C.1. Percentage of PEO Digital Personnel by Occupational Group

Figure C.2. Percentage of PEO Digital Personnel by Grade

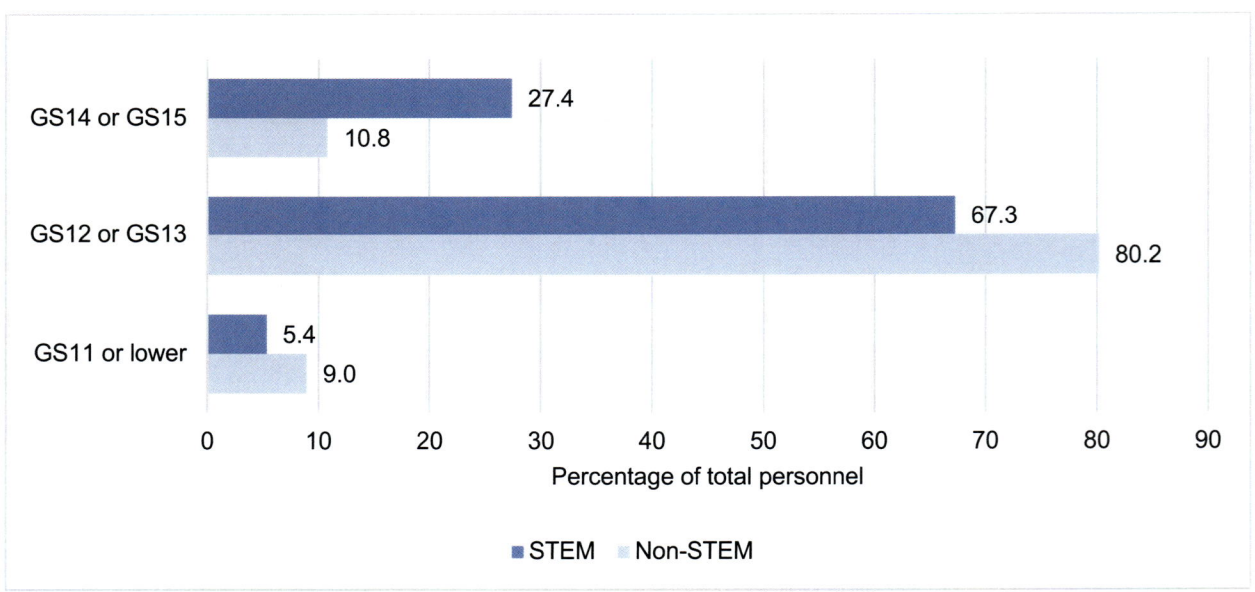

SOURCE: RAND analysis of DCPDS, as of September 2023.
NOTE: Grades across all pay plans are categorized according to the equivalent GS grade.

Overall, STEM personnel in PEO Digital had higher educational attainment than non-STEM personnel. Figure C.3 compares the percentage of STEM and non-STEM personnel by highest degree earned. The percentage of STEM personnel with a graduate degree (including a master's, professional, or doctoral degree) was substantially higher than the percentage for non-STEM personnel, and the percentage of STEM personnel whose highest degree was a bachelor's degree was slightly higher than the percentage for non-STEM personnel.

Figure C.3. Percentage of PEO Digital Personnel by Highest Degree

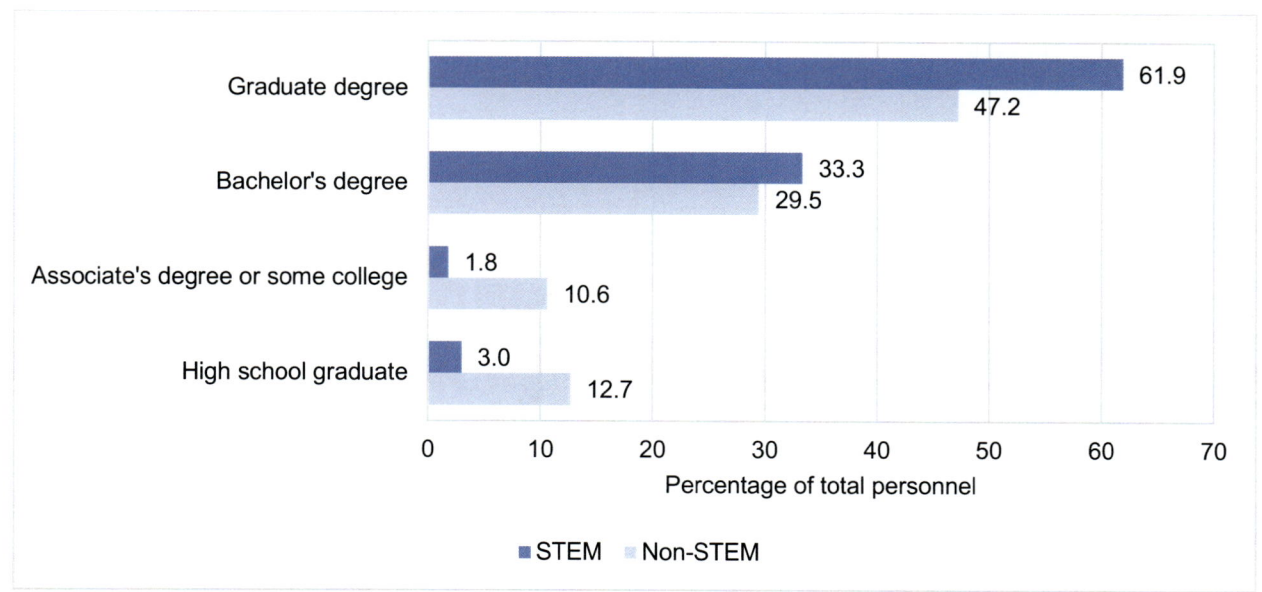

SOURCE: RAND analysis of DCPDS, as of September 2023.

Air Force Futures

We interviewed a total of six Air Force Futures SMEs through six one-on-one interview sessions with individuals at both the staff and the center level. According to one SME, Air Force Futures aims to maintain a flat organization to facilitate cross-center conversations and seamless force design. While they maintain a traditional organizational structure utilizing a chain-of-command flow of authorities, they also developed and maintain a honeycomb-style organizational chart representing the organization's goals of horizontal collaboration across centers.

Table C.1 presents the occupational series code description designations for the Air Force Futures civilian workforce. These are the career fields of civilians working in the organization, as of March 2024. Across this population, approximately 14 percent of Air Force Futures civilians reflect traditional STEM career fields: 19 are operations researchers (1515 series), comprising 13 percent of the organization's civilians; 2 are general engineers (0801 series), comprising 1 percent of the organization's civilians.

Table C.1. Air Force Futures Civilian Positions

Occupational Series Code Description	Frequency	Percentage
Management and Program Analysis (0343)	87	59.59
Miscellaneous Administration and Program (0301)	26	17.81
Operations Research (1515)	19	13.01
Foreign Affairs (0130)	6	4.11
General Engineering (0801)	2	1.37

Occupational Series Code Description	Frequency	Percentage
Security Administration (0080)	2	1.37
Budget Analysis (0560)	1	0.68
Intelligence (0132)	1	0.68
International Relations (0131)	1	0.68
Logistics Management (0346)	1	0.68

SOURCE: Data extracted from the DCPDS, compiled June 3, 2024.

As of March 2024, according to data provided to the RAND team by Air Force Futures, the organization has 29 funded but unfilled civilian positions. Of these unfilled positions, six are in STEM fields: five in operations research and one in data science (1560 series). The rest of the unfilled positions are in management and program analysis (16 positions), miscellaneous administration and program (five positions); foreign affairs (one position), and intelligence (one position). Additionally, Air Force Futures had 11 unfunded civilian positions in Center 1, Concepts and Strategy, as of March 2024. These positions are coded as general staff level roles, with occupational titles to be determined.

Headquarters, Pacific Air Forces

We interviewed a total of eight HQ PACAF SMEs through six interview sessions with individuals working in PACAF/A1 (Manpower, Personnel, and Services), PACAF/A2 (Intelligence Surveillance and Reconnaissance), PACAF/A9 (Analyses, Assessments, and Lessons Learned), the office of the PACAF chief scientist, and the director of staff. Representatives from these offices provided insights and expertise on civilian personnel requirements, utilization, management, hiring, and STEM needs. Participants also provided information on the organizations and missions of the various directorates and staff offices.

HQ PACAF is structured to effectively manage and support its extensive operations across the Indo-Pacific region. The headquarters organization is divided into various directorates and staff offices, each responsible for specific functional areas such as operations, logistics, intelligence, communications, plans, and programs. These directorates work collaboratively to ensure that PACAF's strategic objectives are met and that its forces are prepared to execute their missions. As of September 2023, there were 14 civilian STEM personnel in HQ PACAF comprising 5 percent of 267 total HQ PACAF civilian personnel. STEM personnel were in PACAF/A3/6 (Air and Cyberspace Operations), PACAF A4 (Logistics, Engineering, and Force Protection), PACAF/A9, and the commander's staff.

To contextualize STEM personnel within HQ PACAF, Figure C.4 compares the percentage of individuals in HQ PACAF who were STEM personnel with the percentage in other occupational groups. There were nearly nine times as many individuals in the General Administrative, Clerical, and Office Services group and two times as many individuals in the Social Science, Psychology, and Welfare group as there were STEM civilian personnel. This relatively low percentage of personnel in STEM occupational groups is not unexpected for a MAJCOM headquarters. HQ PACAF was included in our set of case studies for precisely this reason. Given the increasing technical nature of warfare, it may be that there is now unsatisfied demand for STEM in headquarters organizations.

Figure C.4. Percentage of Headquarters, Pacific Air Forces Personnel by Occupational Group

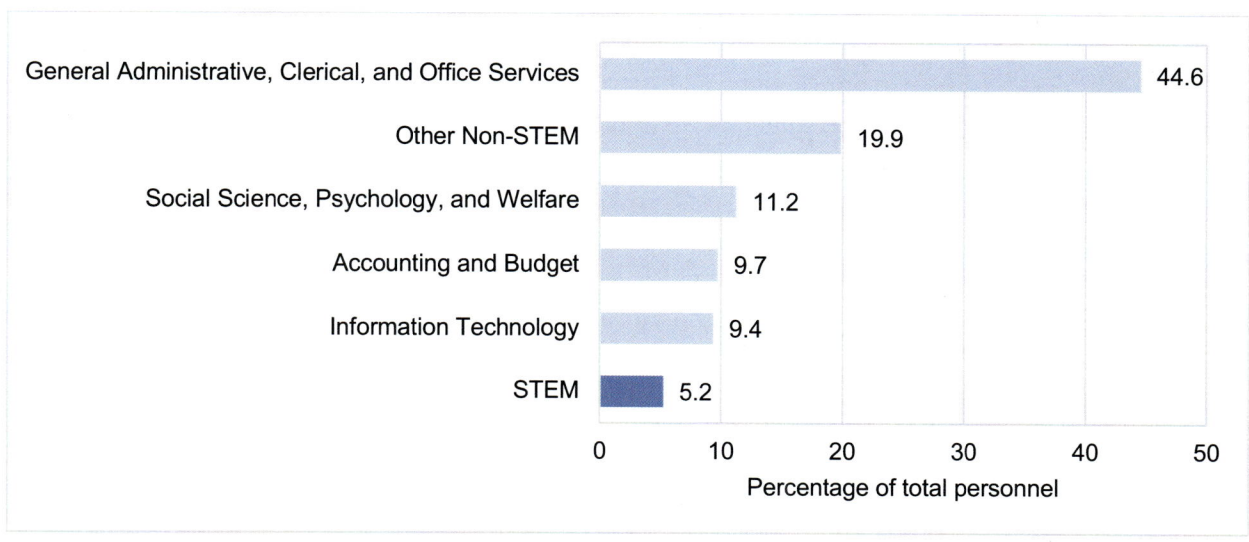

Overall, PACAF STEM personnel were in higher grades than non-STEM personnel. Figure C.5 compares the percentage of STEM and non-STEM civilian personnel by grade. A higher percentage of STEM personnel than non-STEM personnel were in Grades GS-13 through GS-15, and a lower percentage were in Grades GS-12 and below.

Figure C.5. Percentage of Headquarters, Pacific Air Forces Personnel by Grade

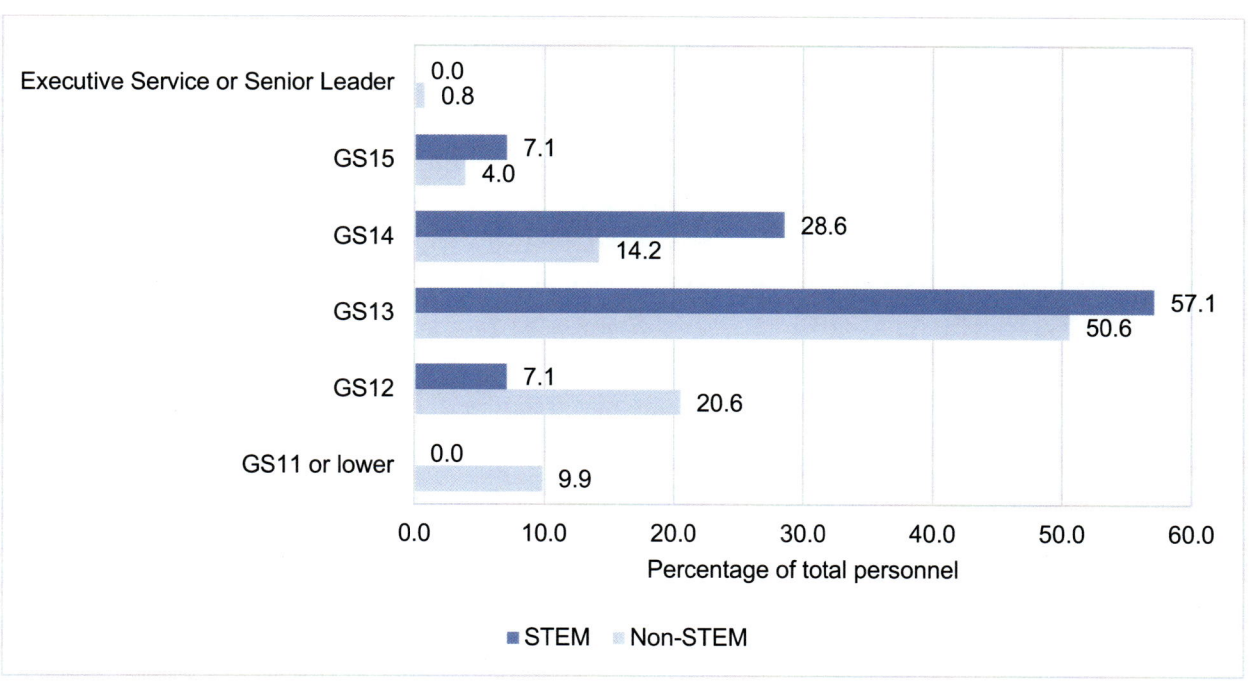

Overall, STEM personnel had higher educational attainment than non-STEM personnel. Figure C.6 compares the percentage of STEM and non-STEM personnel by highest degree earned. The percentage of STEM personnel whose highest degree was a bachelor's was more than twice the percentage of non-STEM personnel, and a higher percentage of STEM personnel had a graduate degree (including a master's, professional, or doctoral degree).

Figure C.6. Percentage of Headquarters, Pacific Air Forces Personnel by Highest Degree

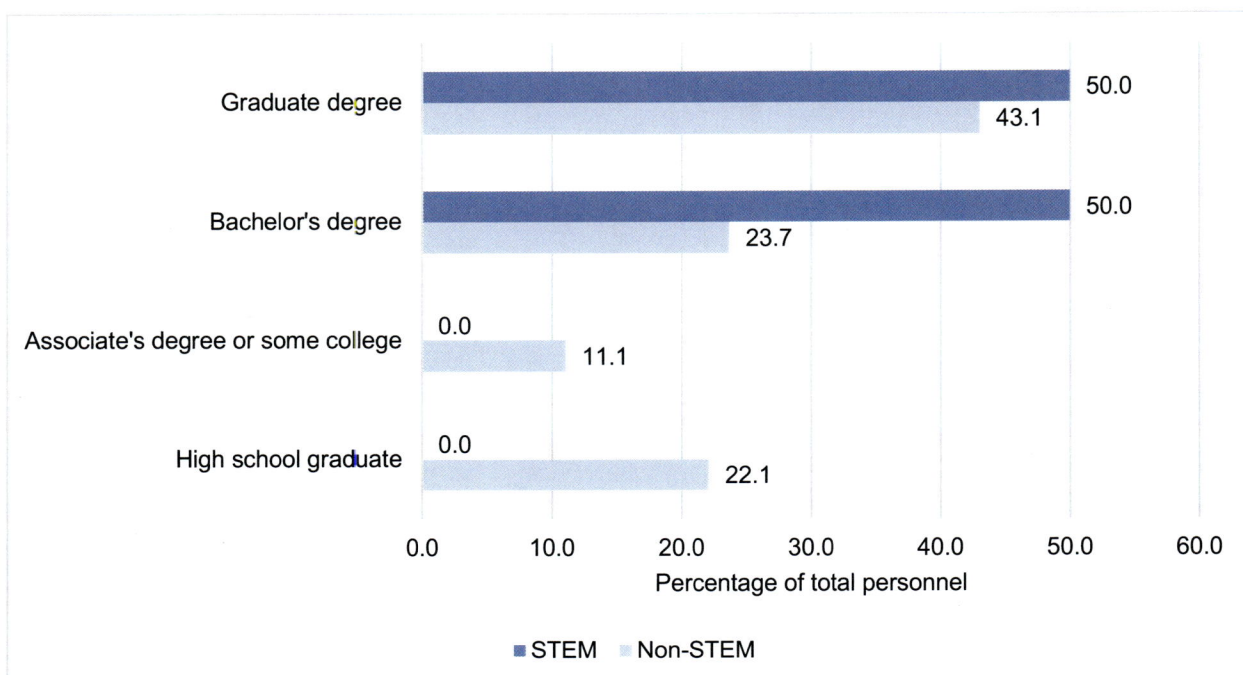

SOURCE: RAND analysis of DCPDS, as of September 2023.

Figure C.7 shows the percentage of HQ PACAF personnel by academic specialty of highest earned degree. Of note is, not all STEM personnel had STEM degrees as their highest degree earned. Specifically, four of the 14 total STEM personnel had a highest degree related to business administration and management or law. These individuals may have been working in management or administrative roles that did not require STEM competencies, or they may have been practicing STEM competencies acquired through experiences rather than by earning their highest degree (e.g., individuals with non-STEM master's degrees may have acquired STEM competencies through STEM undergraduate degrees).

Figure C.7. Percentage of Headquarters, Pacific Air Forces Personnel by Academic Specialty of Highest Degree

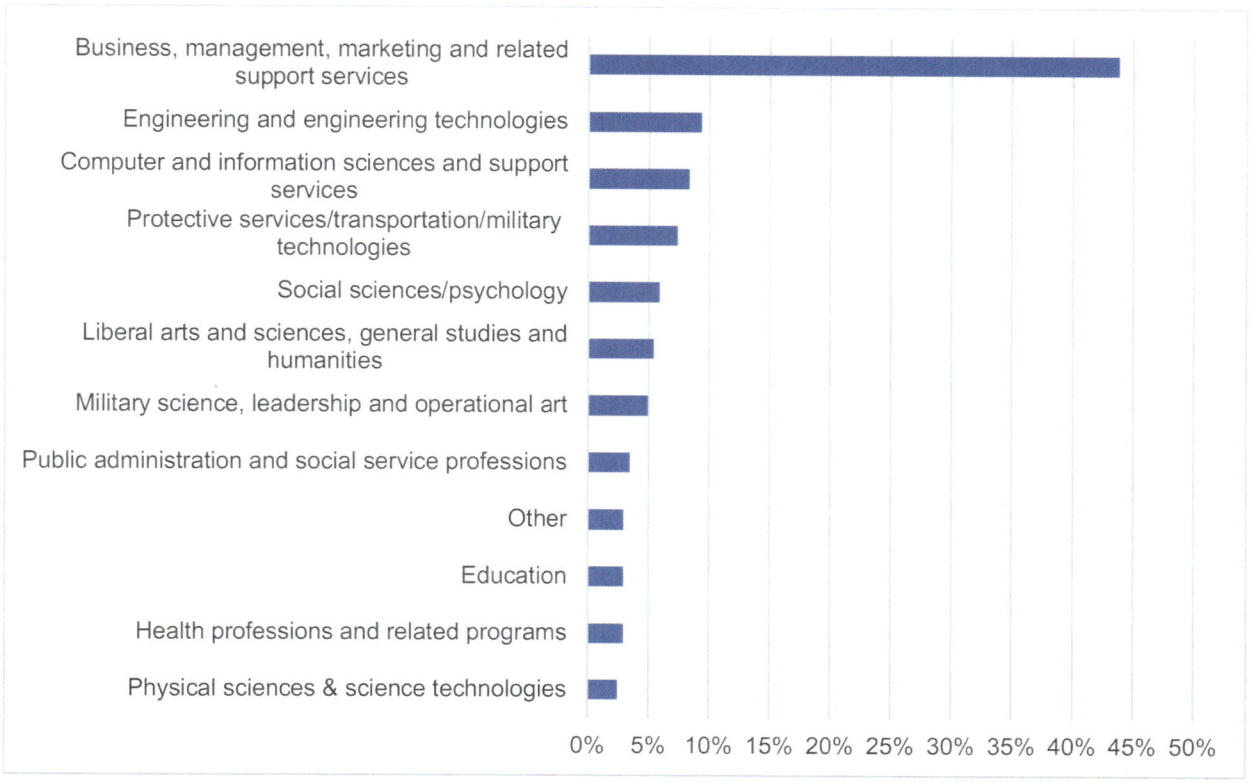

SOURCE: RAND analysis of DCPDS, as of September 2023.

Pilot Surveys for Supply and Demand Assessments of Technical Skills

We developed two surveys to assess STEM personnel competency needs. Individuals from the branch level in PEO Digital responded to the first survey and individuals in Air Force Futures responded to the second survey.

PEO Digital Survey

The PEO Digital survey was developed in an Excel file, which was subsequently shared with and populated by respondents and then returned to the RAND team. The Excel file contained multiple different tabs providing explanations of relevant terms and rating scales for participants, as well as survey questions. As noted in Chapter 5, the PEO Digital survey was developed through an iterative process. The relevant sections of the Excel survey file, reproduced below but adapted for readability, represent the final iterations of survey questions and guidance.

Survey Part 1: Requirements

For the personnel requirements component of the PEO Digital survey, each row contained a critical area of technical expertise in the leftmost column. Some expertise areas were prepopulated by the project team, while others were filled in by participants. Examples of areas of technical expertise in the prepopulated survey include

- digital modeling
- systems engineering
- cybersecurity
- data management.

Participants then addressed a series of eight questions in the proceeding columns for each area of expertise relevant to their branches. The questions posed for each area of expertise and the corresponding response options (selected in the Excel file via a dropdown menu, unless open response) are shown in Table D.1. For definitions of rating options, proficiency gap solutions, and risk scales, see Tables D.3–D.6.

Table D.1. PEO Digital Survey Questions—Part 1: Requirements

Question	Response Options
Is this area of expertise needed in your branch?	Yes
	No
What is the **current** number of employees in your branch with this competency?	(Open response)
What is the number of employees with this competency that your branch **requires**?	(Open response)
Overall, how proficient is your branch in this competency?	Fully proficient
	Mostly proficient
	Moderately proficient
	Significantly deficient
	Critically deficient
Rank the approaches below in order of **effectiveness at addressing a proficiency gap** for the competency (1 = most effective)	Hiring for prior experience
	Hiring for prior education
	Employee education (post-hire)
	On-the-job training
Which of the below would be the greatest challenge in addressing a competency gap?	**Hiring** (i.e., recruitment challenges)
	Authorizations (i.e., too few billets)
	Training (i.e., too time- and resource-intensive to upskill employees)
	No concerns (i.e., no competency gap to be addressed)
What is the level of risk for which your branch will not have the appropriate **level of proficiency** in this competency? (e.g., there are five digital modelers in your branch, but none with expert proficiency.)	Very low risk
	Low risk
	Moderate risk
	High risk
	Very high risk
What is the level of risk for which your branch will not have the appropriate **number of people** with this competency? (e.g., there is one digital modeler in your branch, but you need five given the workload.)	Very low risk
	Low risk
	Moderate risk
	High risk
	Very high risk
Considering any gaps you identified in requirements/areas of expertise, please estimate how many additional personnel (authorizations/FTE) you would need to address those gaps for each source:	(Open response)

Question	Response Options
Additional **civilian** personnel required to close gaps	(Open response)
Additional **military** personnel required to close gaps	(Open response)
Additional **contractors** required to close gaps	(Open response)
Additional **FFRDC/A&AS** personnel required to close gaps	(Open response)

NOTE: A&AS = advisory and assistance services; FTE = full time equivalent.

Survey Part 2: Hiring Challenges

For the hiring challenges component of the PEO Digital survey, potential barriers to the successful hiring of civilian personnel were listed. Respondents were asked to indicate whether their work unit faced that specific hiring challenge by selecting "Yes" or "No." Additional space was given for respondents to name hiring challenges their work units face that were not included in the list provided by the RAND team. A final comment box was also included to give respondents an opportunity to raise any specific thoughts or concerns not covered in the broader survey. The hiring challenges included in the survey are reproduced in Table D.2.

Table D.2. PEO Digital Survey Questions—Part 2: Hiring Challenges

Potential Hiring Challenges	Does your work unit face this hiring challenge?
Uncompetitive wages (primarily compared with the private sector)	Yes/No
Processing/onboarding time (federal application process, background checks, etc.)	Yes/No
Technical expertise gaps (i.e., insufficient number of applicants with a given critical area of technical expertise)	Yes/No
Knowledge level gaps (i.e., education levels, such as M.S. candidates instead of Ph.D. candidates)	Yes/No
Undesirable location (i.e., commute length, geographically remote)	Yes/No
Facilities restrictions (i.e., no/minimal WFH or other flexible work arrangement)	Yes/No
Too few applicants	Yes/No
Criminal record/drug use (i.e., disqualification issues and waiver process)	Yes/No
Lack of childcare options	Yes/No
Insufficient soft skills (i.e., communication, teamwork, writing)	Yes/No
Other (list in next column)	(Open response)
Other (list in next column)	(Open response)

Potential Hiring Challenges	Does your work unit face this hiring challenge?
Do you have additional comments? Enter any other thoughts you wish to share regarding workforce requirements, DoD civilian and other personnel matters, competency gaps and other challenges, future needs, or measures (e.g., on the job training) taken to ensure your work unit meets its objectives.	(Open response)

NOTE: WFH = work from home.

Definitions and Rating Scales

To ensure that respondents understood the nature of the questions being asked, terms were defined and rating scales provided for proficiency and personnel risks. The Excel file containing the survey included tabs with explanations of relevant terms utilized in the file. Those definitions and scale explanations are reproduced in Tables D.3–D.6.

Table D.3. Relevant Survey Definitions

Term	Definition
Prior experience	Refers to the professional work experience an individual has accumulated before being hired into their current position. This includes full-time, part-time, temporary, or contract roles held in any industry or sector that provided relevant skills, knowledge, or expertise applicable to the competency or area of specialization. This definition excludes any experience gained **in the current** position.
Prior education	Refers to the formal educational attainment of an individual that includes the completion of a bachelor's degree or higher, including master's and doctoral degrees (e.g., Ph.D.).
Post-hire education	Refers to the coursework or formal education programs an individual completes while working in their current position.
On-the-job training	Refers to the training an individual receives while working in their current position or in relevant task assignments or rotations. On-the-job training can include a variety of other formats such as mentoring by more experienced colleagues, hands-on practice, and shadowing.
Professional development	Refers to additional attendance at training programs, instructional sessions, workshops, conferences, seminars, and self-directed learning activities.
None/basic	Individual requires no prior experience with this specific technical area or can apply the competency in simple situations with extensive guidance.
Intermediate	Individual is required to apply the competency in typical situations with occasional guidance for complex situations.
Expert	Individual is required to apply the competency independently in complex situations; qualified to serve as an SME and advise others.

Table D.4. Proficiency Scale

Proficiency Level	Definition
Fully proficient	The proficiency level across the branch or unit fully meets the needs with no deficiencies.
Mostly proficient	Proficiency level is generally adequate, with only minor deficiencies that rarely impact overall mission performance.
Moderately proficient	There are noticeable deficiencies in proficiency that sometimes affect mission performance and could pose challenges to operational effectiveness.
Significantly deficient	There are significant deficiencies in proficiency that frequently impact performance and hinder the achievement of operational goals.
Critically deficient	The proficiency level is severely inadequate, consistently impeding mission performance and preventing the branch or unit from fulfilling its objectives.

Survey Question: Overall, how proficient is your branch in this competency?

Table D.5. Risk Scale: Level of Proficiency

*Survey question: What is the level of risk that your branch will not have the appropriate **level of proficiency** in this competency? (e.g., there are five digital modelers in your branch, but none with expert proficiency.)*

Risk Level	Definition
Very low risk	There is minimal chance that your branch will not have the appropriate level of proficiency in this competency in the future.
Low risk	There is a slight risk that your branch will not have the appropriate level of proficiency in this competency in the future. However, minor changes or interventions can likely prevent any significant deficiency.
Moderate risk	There is a noticeable risk that your branch will not have the appropriate level of proficiency in this competency in the future. Proactive measures may be required to address potential deficiencies.
High risk	There is a significant risk that your branch will not have the appropriate level of proficiency in this competency in the future. Immediate and substantial interventions are necessary to mitigate this risk.
Very high risk	There is an extremely high risk that your branch will not have the appropriate level of proficiency in this competency in the future. Urgent and comprehensive actions are imperative to prevent severe deficiencies.

Table D.6. Risk Scale: Number of Personnel

*Survey question: What is the level of risk that your branch will not have the appropriate **number of people** in this competency? (e.g., there are five digital modelers in your branch, but none with expert proficiency.)*

Risk Level	Definition
Very low risk	There is minimal chance that your branch will not have the appropriate number of individuals with this competency in the future.
Low risk	There is a slight risk that your branch will not have the appropriate number of individuals with this competency in the future. However, minor changes or interventions can likely prevent any significant deficiency.
Moderate risk	There is a noticeable risk that your branch will not have the appropriate number of individuals with this competency in the future. Proactive measures may be required to address potential deficiencies.
High risk	There is a significant risk that your branch will not have the appropriate number of individuals with this competency in the future. Immediate and substantial interventions are necessary to mitigate this risk.
Very high risk	There is an extremely high risk that your branch will not have the appropriate number of individuals with this competency in the future. Urgent and comprehensive actions are imperative to prevent severe deficiencies.

Air Force Futures Survey

Table D.7 shows the Air Force Futures survey. Respondents were asked to rate the characteristics in the survey using a value scale of 1 to 5, with 1 indicating that the characteristic has a low value and 5 indicating that the characteristic has a high value. Respondents were prompted to explain their rationale for each characteristic. Additionally, the survey included an open-ended narrative section that prompted respondents as follows: "In addition to weighting characteristics above, please generally describe your ideal candidate for hiring a new civilian into Air Force Futures."

Table D.7. Air Force Futures Survey: Valuing Characteristics in Civilian Hires

Survey Question: When considering the current and future needs of Air Force Futures, what characteristics in a civilian hire would be most important to help Air Force Futures meet mission objectives? Consider the six characteristics below. Use the drop-down menu to indicate how much you value this characteristic in a civilian hire to meet Air Force Futures current needs and to meet Air Force Futures future needs. Please explain your rationale.

Characteristic	Value—Current	Value—Future	Explanation
Academic degree—bachelor's with STEM focus			
Academic degree—bachelor's with non-STEM focus			
Academic degree—master's or Ph.D. with STEM focus			
Academic degree—master's or Ph.D. with non-STEM focus			

Survey Question: When considering the current and future needs of Air Force Futures, what characteristics in a civilian hire would be most important to help Air Force Futures meet mission objectives? Consider the six characteristics below. Use the drop-down menu to indicate how much you value this characteristic in a civilian hire to meet Air Force Futures current needs and to meet Air Force Futures future needs. Please explain your rationale.

Technical expertise (i.e., expertise in AI, modeling, data analytics, etc.)

DoD experience (i.e., work experience in a DoD organization)

Private-sector experience (i.e., work experience in private industry)

NCR experience (i.e., work experience in the national capital region)

Development of STEM Pipelines and Capabilities

To evaluate how the DAF develops and invests in civilian STEM personnel, we explored the educational and professional development initiatives targeted at the DAF's civilian STEM pipelines and capabilities, which are detailed in this appendix. Many of these initiatives are available not only to civilians but also to active duty, guard, and reserve personnel, with some overlap between offerings for uniformed and nonuniformed personnel. Additionally, there are numerous other developmental programs and initiatives that, while not directly focused on STEM, may indirectly support STEM competencies.

This appendix does not aim to comprehensively document or evaluate all the programs and initiatives within the DAF ecosystem. Instead, our goal is to broadly describe some of the key professional pathways for civilians in STEM career fields and how these pathways can support the broader development of STEM competencies.

Development of Student STEM Pipelines

The DAF provides a variety of initiatives designed to foster early interest and proficiency in STEM fields among students. From high school to graduate school, these programs aim to identify and nurture STEM talent while encouraging scholars to consider careers within the DAF. Many of these initiatives are specifically targeted at integrating STEM expertise into the DAF civilian workforce.

The primary objectives of these programs include generating interest in STEM careers, leveraging the potential of innovation hubs, and providing hands-on STEM experience. Through these efforts, the DAF seeks to build a pipeline of skilled individuals equipped to meet the technological and scientific challenges of the future. Table E.1 summarizes such programs.

Table E.1. Student STEM Pipeline Programs

Program	Target Audience	Description
Leadership Experiences Growing Apprenticeships Committed to Youth (LEGACY)	Middle and high school students	Three-phase initiative with week-long camps for middle schoolers and paid summer internships for high schoolers
Strategic Ohio Council for Higher Education (SOCHE)	High school and college students	Enhances research capabilities and professional relationships with AFRL technical directorates.
Wright Scholar Research Assistant Program	High school juniors and seniors	Full-time, paid summer internship with AFRL scientists and engineers
Awards to Stimulate and Support Undergraduate Research Experiences (ASSURE)	Undergraduate students	Semester-long, stipend-paid full-time work at DoD laboratories
AFRL Scholars Program	Undergraduate and graduate students	Stipend-paid summer internships with AFRL scientists and engineers
Student Research Program	Undergraduate and graduate students	Year-long internships at both AFIT and AFRL
Advanced Course in Engineering (ACE) Internship Program	Undergraduate students, Reserve Officers Training Corps (ROTC) cadets, select active-duty personnel	Stipend-paid summer internship focusing on cyber warfighting and leadership
Research Laboratory Information Directorate Summer Internship in Rome, New York	Undergraduate and graduate students	Stipend-paid summer internships focused on computer-centered research
Minority Leaders Research Collaboration Program (ML-RCP)	Undergraduate and graduate students	Stipend-paid summer internships for minority students
AFRL STEM Student Employment Program	Undergraduate and graduate students	Full- or part-time paid internships during the semester
Oak Ridge Institute for Science and Education (ORISE)	Undergraduate and graduate students, recent graduates, postdoctoral students, faculty	DoD innovation hub offering a large portfolio of internships, fellowships, and research funding such as the Repperger Research Internship
National Defense Science and Engineering Graduate Fellowship Program	Graduate students	Funding and support for advanced degrees in critical STEM fields
University Research and Engagement Program (UREP)	Undergraduate, graduate, and postdoctoral students	Funds research proposals from graduate student–faculty teams in space related fields
Defense Associated Graduate Research Innovators Program	Graduate students and faculty	Funds research proposals from graduate student–faculty teams in engineering related fields

Program	Target Audience	Description
AFRL Scholars Professionals	Early career graduates	Year-long full-time research contracts at Kirkland Air Force Base

SOURCES: Information compiled from Department of the Air Force Manual 36-142, *Civilian Career Field Management and Centrally Managed Programs*, Secretary of the Air Force, October 4, 2022; Air Force Research Laboratory, "Internships and Scholarships," webpage, undated-c; Air Force Research Laboratory, "AFRL Scholars," undated-b; Caroline M. Miller, "Military Department Personnel Posture Hearing," testimony submitted to the Subcommittee on Military Personnel, Committee on Armed Services, United States House of Representatives: Military Department Personnel Programs, March 9, 2023; Deputy Assistant of the Air Force (Science, Technology and Engineering), *Civilian Career Development Guide for DAF Scientist & Engineer Career Field Scientists, Engineers & Technicians*, Version 1.0, June 12, 2023; Strategic Ohio Council for Higher Education, homepage, undated; and The Griffiss Institute, homepage, undated.

NOTE: The student STEM pipeline programs listed in this table are not comprehensive but rather represent a selection of existing opportunities.

Programs to Recruit Student STEM Talent

In addition to sustained investments in building STEM pipelines through funding student scholarship, the DAF also operates programs to recruit STEM talent into the civilian workforce. For example, the Premier College Intern Program (PCIP) offers college students STEM-focused internships that seamlessly transition into permanent or entry-level positions within the DAF. Upon completing internship requirements, the individual may be eligible for a full-time DAF position. In 2023, the PCIP intern retention rate was 86 percent.[1]

For college graduates, the DAF offers three formal career development programs:

- The COPPER CAP Program is a four-year initiative that trains college graduates as contract specialists within the Air Force, offering career development opportunities such as formal education and on-the-job training.[2]
- The PALACE Acquire Intern[3] is a three-year Air Force program designed for college graduates from STEM fields. Participants access benefits such as recruitment bonuses, student loan reimbursement, paid relocation, and funded graduate degrees provided they meet work requirements. The goal of PALACE Acquire is to develop these individuals into professional Air Force Civilian Scientists and Engineers.
- The AFRL Scholars Professionals program offers early career graduates the opportunity to work full-time for one year, renewable annually up to three years at Kirtland Air Force Base, in STEM, business administration, and cybersecurity.

The AFRL Science and Technology Fellowship Program offers full-time fellowships for post-doctoral students and senior research associates to conduct self-directed research at Air Force and

---

[1] Caroline M. Miller, "Military Department Personnel Posture Hearing," testimony submitted to the Subcommittee on Military Personnel, Committee on Armed Services, United States House of Representatives: Military Department Personnel Programs, March 9, 2023.

[2] Deputy Assistant of the Air Force (Science, Technology and Engineering), 2023.

[3] DAF Manual 36-142, *Civilian Career Field Management and Centrally Managed Programs*, Secretary of the Air Force, October 4, 2022.

Space Force laboratories. Similarly, the Air Force Office of Scientific Research Summer Faculty Fellowship Program provides eight- to 12-week residencies for full-time science, mathematics, and engineering faculty at Air Force and Space Force laboratories.[4]

Initiatives Targeted at the Development of Civilian STEM Education

In addition to building STEM pipelines through funding scholarships and internships, the DAF employs programs such as tuition support and reimbursement, comprehensive educational opportunities, and numerous certification and on-the-job training programs to develop its civilian STEM workforce. Civilians can learn about these opportunities through their supervisors and development teams, through networks such as the Digital Transformation Office's Digital Transformation Network, and through resources such as the *Civilian Career Development Guide for DAF Scientist & Engineer Career Field Scientists, Engineers & Technicians*.[5]

Tuition Support and Reimbursement

The primary form of tuition support for civilian personnel is the Civilian Tuition Assistance Program. This program funds full-time and part-time degree and certification programs.[6] In addition, the SMART Scholarship-for-Service Program provides funding for B.S. through Ph.D. programs in STEM disciplines. While SMART primarily supports active duty, guard, and reserve personnel, it is also available to current government employees.[7]

Educational Opportunities

The educational ecosystem of the DAF is extensive, offering civilians opportunities in federal service academies, DoD-affiliated universities, and educational programs managed by both DoD and the DAF. Prominent institutions include federal service academies such as the United States Military Academy, Naval Academy, Air Force Academy, Coast Guard Academy, and Merchant Marine Academy. While these institutions have the primary goal of educating and commissioning officers, civilian personnel can teach at service academies and access STEM-related professional development opportunities while there.

In addition, specialized institutions such as the Naval Postgraduate School, Air University, AFIT, and the Uniformed Services University of the Health Sciences provide advanced education and training to eligible U.S. military students, international students, DoD civilian employees, and a limited number of defense contractors.

---

[4] AFRL, undated-c.

[5] Deputy Assistant of the Air Force (Science, Technology and Engineering), 2023.

[6] Miller, 2023.

[7] Deputy Assistant of the Air Force (Science, Technology and Engineering), 2023.

There are also government-run "corporate" universities such as DAU, which operates under DoD within the Office of the Under Secretary of Defense for Acquisition, Technology, and Logistics.[8] Finally, civilian universities such as MIT, Stanford, Johns Hopkins, Georgia Tech, and the University of Southern California play a significant role in defense-related research and development projects,[9] contributing to advancements in defense technologies and strategic studies.[10] Civilian personnel also have opportunities to obtain education through tuition assistance at these universities, although they must go through the same application processes as typical students.

In the sections below, we highlight educational opportunities for civilians at two key DAF-affiliated organizations: Air University and AFIT.

Air University

Located at Maxwell Air Force Base in Alabama, Air University offers professional military education, academic degree programs, and continuing education to prepare military and civilian USAF personnel for leadership roles and professional skills enhancement. For civilians in STEM career fields, Air University offers access to an online master's degree program providing flexible, tuition-free education in concentrations such as Joint Warfare, Leadership, Nuclear Weapons, and Operational Warfare.[11]

Air Force Institute of Technology

AFIT offers an array of opportunities for the STEM workforce, which are aimed at enhancing technical proficiency and leadership capabilities across various disciplines. AFIT's Graduate School of Engineering and Management offers 265 research-based STEM master's degree programs, 15 Ph.D. programs, and 16 graduate certificate programs, with typical enrollment exceeding 650 in-residence students and about 400 in distance-learning programs. In 2022, 26 percent of AFIT graduate students were civilians.[12] AFIT also offers professional continuing education across three schools: the Civil Engineering School, the School of Strategic Force Studies, and the School of Systems in Logistics.

AFIT has undergone recent efforts to tailor its programs to emerging digital needs. In October 2023, it established the Digital Innovation & Integration Center of Excellence to enhance digital excellence.[13] The center aims to increase educational excellence, research and technology transfer, consulting, and best practices in digital technologies. To do so, it actively collaborates with the Digital

---

[8] U.S. Code, Title 10, Armed Forces; Subtitle A, General Military Law; Part II, Personnel, Chapter 87, Defense Acquisition Workforce; Subchapter IV, Education and Training.

[9] For example, MIT and DAF partnered to create the AI Accelerator, a state-of-the-art pipeline for AI technology in February 2019. This partnership includes interdisciplinary projects involving MIT faculty and Air Force personnel to advance AI research in areas such as weather modeling, training optimization, and decisionmaking autonomy.

[10] Office of the Under Secretary of Defense for Research and Engineering, "Federally Funded Research and Development Centers (FFRDC) and University Affiliated Research Centers (UARC)," webpage, undated.

[11] Air University, "Information and Facts 2022 & 2023," fact sheet, 2023.

[12] AFIT, "By the Numbers FY 22," fact sheet, 2022.

[13] AFIT, "Digital Innovation & Integration Center of Excellence," webpage, undated.

Transformation Office and works with administration to increase offerings of graduate, professional, and continuing education classes in digital technologies.[14]

Certifications

The USAF provides its civilians, military personnel, and contractors with numerous opportunities for short-term certifications. These programs are especially beneficial for delivering short, focused training that enables rapid reskilling and upskilling. Such programs enable personnel to acquire new competencies for transitioning into emerging roles (reskilling) or to deepen their expertise within their current domain (upskilling), with the goal of ensuring that the USAF remains agile and capable of meeting the demands of modern technological and operational environments. Table E.2 summarizes some of the main organizations that offer these certifications.

Table E.2. Core Certification Institutions

Institution	Ownership	Focus Areas
AFIT certification programs	USAF	Advanced education and technical training for DAF personnel in engineering, logistics, and management
Computing Technology Industry Association (CompTIA) programs	Private	Information technology, including cybersecurity, networking, and technical support; supports cyber personnel to meet DoD Manual 8140.03 requirements
DAU certification programs	DoD	Acquisition, technology, and logistics training for the defense workforce
Digital University certification programs	USAF	Digital skills and technologies, including data science, artificial intelligence, and cloud computing

SOURCE: Adapted from Deputy Assistant of the AirForce (Science, Technology and Engineering), 2023.
NOTE: These certification programs are not exhaustive; they aim to represent a selection of the principal institutions that responsible for providing STEM-related certifications to personnel affiliated with USAF.

Civilian Professional Development Programs to Enhance STEM Skills

The DAF enhances its STEM talent through various on-the-job training opportunities with technical experts both inside and outside the service. One prominent type of on-the-job training is career broadening.[15] This aims to expand an employee's knowledge beyond their functional or technical area in order to enhance their enterprise awareness and leadership abilities. Career broadening can be achieved through the centrally managed Career Broadening Program or by

---

[14] For example, AFIT now offers a graduate-level digital engineering program, developed and funded through a partnership between Hanscom Air Force Base, University of Massachusetts, Lowell, the Massachusetts Military Asset and Security Strategy Task Force, and MassDevelopment. The program consists of four semester-long courses focused on DE methods, models, and strategies. For more see Jessica Casserly, "First Cohort Graduates Digital Engineering Program, Gains Vital Skills," press release, 66th Air Base Group Public Affairs, January 24, 2024.

[15] Deputy Assistant of the Air Force (Science, Technology and Engineering), 2023.

undertaking assignments at different levels (e.g., flight, squadron, branch, division) and across various technical areas.

Civilians also have access to fellowships that provide technical and operational experience. For example, the DAF Research Fellowship Program is a 10- to 18-month program for select field-grade officers and civilian counterparts.[16] This program offers in-depth education in national security policy through assignments at distinguished civilian institutes or key government agencies.

In addition, civilians can participate in projects sponsored by AFWERX, the innovation branch of AFRL. One notable AFWERX program is Spark, which fosters an innovation culture by connecting diverse communities and accelerating the adoption of promising technologies. AFWERX also sponsors the AFWERX Fellowship, a four-month innovation training program that immerses both uniformed and nonuniformed personnel in the AFRL innovation ecosystem and provides participants with key responsibilities and a structured curriculum on various innovation topics.[17]

On-the-job training is also available outside of the DAF. For example, Education with Industry is a ten-month USAF-run initiative that provides officers and civilians the chance to intern at top-tier public and private-sector companies.[18] Another popular opportunity is the Engineer and Scientist Exchange Program, which allows personnel to work with foreign counterparts on international cooperative research, development, testing, and evaluation through full-time, on-site assignments.[19] Finally, the AFRL Regional Network Program offers AFRL personnel up to six months to work on high-risk research with regional private, academic, and military network partners.[20]

Workshops, Training Series, and Holistic Professional Development

The DAF offers a wide range of workshops, training series, and other initiatives for its civilian personnel. One notable effort is the expansion of occupational training in emerging and rapidly evolving STEM fields, such as DE and MBSE, through the Digital Facilitators Academy.[21]

The academy aims to equip both uniformed and nonuniformed personnel with essential, cross-functional digital skills. The Digital Facilitators Academy's training series starts with Facilitation 101, which introduces new leaders from programs such as PALACE Acquire and PCIP to digital facilitation techniques. Participants engage in interactive workshops designed to enhance their ability to lead virtual meetings effectively by focusing on collaboration and actionable insights. Workshop topics include mastering facilitation, strategic meeting organization, and storytelling with data, all aimed at boosting career potential and accelerating transformation within the Air Force.

In addition, the academy offers Continuous Learning Education points and a Digital Badge upon completion. Such a badge provides a digital credentialing framework that allows both uniformed and

---

[16] Deputy Assistant of the Air Force (Science, Technology and Engineering), 2023.

[17] AFWERX, *2023 Annual Report*, 2023.

[18] DAF Manual 36-142, 2022.

[19] DAF Manual 36-142, 2022.

[20] AFRL, "AFRL Regional Network Overview," webpage, undated-a.

[21] Digital Transformation Office, "Learn How to Drive Successful Online Collaboration with the DAF Digital Facilitators Academy," webpage, November 20, 2023.

nonuniformed personnel to showcase their knowledge, skills, and experience; it can be retained indefinitely and transferred to nonmilitary sectors.

Targeted Professional Development in STEM Areas

In some STEM areas, such as nuclear technologies, the USAF has implemented targeted professional development programs for specific occupations. For example, DAF Instruction (DAFI) 13-504, *Nuclear Mission Professional Development*, outlines a comprehensive approach to developing civilian and uniformed nuclear professionals.[22] The document details the proficiency levels and behaviors required for each key nuclear occupation competency and specifies the necessary technical instruction, nuclear-specific professional continuing education, and experiential learning needed to achieve these goals.

The nuclear field also has mechanisms to ensure that key billets are filled.[23] DAF Instruction 13-504 includes specific communication protocols between cross-functional and functional authorities to ensure that manpower forecasts, human capital requirements, and learning needs for each Air Force specialty code and occupational series are accurate. This enables development teams to effectively grow professionals based on specific learning needs.

---

[22] DAFI 13-504, *Nuclear Mission Professional Development*, Secretary of the Air Force, November 23, 2021.

[23] The Nuclear Key Billet Program supports nuclear deterrence, nuclear acquisition, and nuclear security by identifying where specialized nuclear talent is needed and thereby improving career path management and policy implementation. In addition, the Nuclear Civilian Billet Development initiative focuses on filling nuclear civilian positions and providing professional development opportunities for these roles. The Nuclear Command, Control, and Communications Talent Management initiative ensures that NC3 positions are staffed with qualified personnel.

Broad Civilian Workforce Challenges in the Department of the Air Force

In this appendix, we provide an overview of broad challenges that affect how the DAF can track and manage its STEM workforce.

Prior Civilian STEM Personnel Research

Workforce planning research has documented several challenges the DAF faces in its efforts to hire individuals with requisite technical competencies. These include "a small talent pool with the desired skills and credentials for some occupations, lower pay relative to the private sector, lengthy hiring timelines, challenges with hiring in remote or overseas locations, and a lack of effective marketing and recruiting at the local level."[1] Such challenges exist in a hiring environment in which needs for high-quality technical talent—both in the DAF and the broader U.S. labor market—are expected to grow, especially in such STEM areas as data science and mathematics.[2]

Like other federal entities, the DAF has certain hiring authorities that allow it to expedite the recruitment process for personnel with critical skills and to offer employment packages that bring salary and/or promotion opportunities closer to the private sector.[3] Still, given some of the unique aspects of working for the DAF (e.g., obtaining a security clearance, less flexible work-from-home opportunities due to requirements to work in secure facilities), compensation or promotion opportunities that mirror the private sector are not enough to level the playing field.

Literature on this topic also details a structural challenge associated with hiring civilians in the federal government, including for research roles in the DAF. Specifically, the way in which OPM classifies jobs according to an occupational series—with each one entailing a unique set of requirements, such as type of education and degree and salary range—"can create obstacles to hiring civilians who have the technical skills needed but perhaps do not meet the exact degree requirements specified by OPM."[4] A recent review of RAND literature on talent management in DoD arrived at a

---

[1] Keller et al., 2023, p. v.

[2] Keller et al., 2023, p. 1.

[3] Keller et al., 2023, pp. 7–22.

[4] Kirsten M. Keller, Maria C. Lytell, and Shreyas Bharadwaj, *Personnel Needs for Department of the Air Force Digital Talent: A Case Study of Software Factories*, RAND Corporation, RR-A550-1, 2022, p. 12.

similar conclusion, stating that the department "struggles with defining required capabilities and job classifications" for important knowledge-based roles, such as data science and cyber.[5]

In practical terms, DAF managers looking to hire civilians are occasionally faced with a choice between listing a position in an occupation that pays more but provides less flexibility in degree type or that offers more flexibility in degree requirements but pays less.[6] Moreover, while supervisors of research units have noted that some gaps do exist between currently authorized billets and those required to meet mission (whether currently or in the future), precise personnel needs appear to be unknown.[7] Prior work suggests that determining personnel needs is a challenging task for a number of reasons, ranging from the difficulty of predicting growth of needs in particular specialties, imprecise competency models (i.e., models that define required technical skills but not proficiency levels), and a lack of coordination across DAF career fields, which limits system-wide understandings of needs.[8] In other words, the DAF may be struggling with civilian hiring not only from the perspective of incentives and job requirement constraints, but also from a lack of insight into what its true personnel needs are.

Different Personnel Systems Make Managing STEM Personnel Challenging

Managing STEM personnel in the DAF is complicated by the variety of authorities and programs available for hiring and setting pay for civilian STEM personnel. Different organizations at a single Air Force base may leverage these tools differently, giving some units an advantage in attracting and retaining high-demand STEM workers. This can impede a senior manager's ability to allocate STEM talent where it is most needed.

The traditional federal personnel system governed by Title 5 of the U.S. Code, provides managers with limited flexibility in hiring and setting pay for civilian employees. White-collar federal employees, including STEM personnel, are hired under Title 5 and paid under the GS classification and pay system, which establishes 15 pay grades and ten pay steps within each grade. Step increases are based on acceptable performance and longevity, while promotions to higher grades require competition or completion of a career-ladder position with promotion potential.[9] Title 5 mandates "equal pay for substantially equal work" and allows pay variations only in proportion to "substantial differences in the difficulty, responsibility and qualifications requirement of the work performed and the contributions of employees to efficiency and economy of the service."[10] Consequently, the GS system limits managers' ability to offer higher pay rate increases to recruit or retain individuals in high-demand occupations. In

[5] Laura Werber, *Talent Management for U.S. Department of Defense Knowledge Workers: What Does RAND Corporation Research Tell Us?* RAND Corporation, RR-A950-1, 2021, p. viii.

[6] Keller, Lytell, and Bharadwaj, 2022, p. 12.

[7] Keller, Lytell, and Bharadwaj, 2022, p. 13.

[8] Werber, 2021, pp. 7–9.

[9] Groeber et al., 2021.

[10] U.S. Code, Title 5, Section 5101, Purpose.

addition, Title 5 requires a competitive hiring process open to all applicants, which managers often indicate is rigid and complex, while expressing a need for more flexibility.[11]

Under the traditional system, managers have some tools to offer incentives to high-demand workers and streamline hiring. Recruitment, relocation, and retention bonuses can be offered under specific conditions: recruitment bonuses for positions difficult to fill, relocation bonuses for positions in challenging geographic areas, and retention bonuses for essential employees at risk of leaving. These bonuses require adherence to federal law criteria and agreements between the employee and agency. Funding for these incentives typically comes from a base's or activity's civilian personnel budget. To ease hiring, direct hire authority has been granted for some positions, eliminating competitive rating and ranking procedures and allowing faster recruitment. For example, the 2019 National Defense Authorization Act granted direct hire authority to DoD for any science, technology, or engineering position.[12] Several programs have been created to provide flexibility in pay and hiring for specific civilian positions. As discussed earlier in the report, these include AcqDemo, Lab Demo, DCIPS, CES, and PCIP.

Table F.1 shows the types of personnel covered by each program. AcqDemo, Lab Demo, and DCIPS use pay bands, which create an alternative pay grade structure by grouping two or more GS grades together. One pay band has the salary range of several GS grades, which allows employees to receive larger salary increases without competing for a promotion.[13] AcqDemo and Lab Demo also include contribution-based compensation systems, which set pay increases based on an employee's contributions in specific areas. CES allows DoD to make an adjustment to permanent pay that targets workers in a specific cyber role, grade, and location where high turnover or difficulty filling vacancies exists.[14] PCIP is a summer internship that allows STEM degree students to enter a two- to three-year training and development track, culminating in a permanent position. It includes performance-based promotions and recruitment bonuses for selected positions upon full-time hire.[15]

Table F.1. Types of Personnel Covered by Selected Programs

Program	Personnel Covered
AcqDemo	Certain DoD acquisition personnel
Lab Demo	Certain AFRL personnel
DCIPS	DoD intelligence community employees
DoD CES	DoD cybersecurity employees
PCIP	Students pursuing a bachelor or master of science degree in STEM-related fields

SOURCES: Adapted from Keller et al., 2023; Knapp et al., 2021; Air Force Civilian Service, Science & Engineering, webpage, undated.

---

[11] OPM, "Competitive Hiring," webpage, undated-b; Groeber et al., 2020.

[12] Keller et al., 2023; Groeber et al., 2020.

[13] Groeber et al., 2021.

[14] Knapp et al., 2021.

[15] Air Force Civilian Service, "Science & Engineering," webpage, undated.

These programs create different sets of tools for hiring and retaining STEM workers. Different organizations at a single location may have varying abilities to hire and retain the same STEM worker. For example, pay bands might make it easier for AFRL to attract an engineer than it would for an organization under the traditional GS system.[16] PCIP could also allow organizations linked to the internship program to bring on an engineer outside the traditional hiring process. Conversely, organizations using the traditional system might face constraints in providing hiring and retention bonuses due to funding limitations from base or activity funds. While these special tools can aid in hiring and retention for organizations that use them, they may impede a senior manager's or a human resource professional's ability to ensure that personnel fill positions where they are most needed.

---

[16] For additional discussion and insights related to compensation and pay plans, see Jessie Coe, Maria C. Lytell, Christina Panis, and William Shelton, *Demographic Diversity of the Science, Technology, Engineering, and Mathematics (STEM) Workforce in the U.S. Department of Defense: Analysis of Compensation and Employment Outcomes*, RAND Corporation, RR-A1480-1, 2023; Jennifer Lamping Lewis, Laura Werber, Cameron Wright, Irina Elena Danescu, Jessica Hwang, and Lindsay Daugherty, *2016 Assessment of the Civilian Acquisition Workforce Personnel Demonstration Project*, RAND Corporation, RR-1783-OSD, 2017.

Abbreviations

AcqDemo	Department of Defense Civilian Acquisition Workforce Personnel Demonstration Project
AF/ST	Air Force Chief Scientist
AFIT	Air Force Institute of Technology
AFLCMC/HB	Air Force Life Cycle Management Center, Digital Directorate (aka PEO Digital)
AFPC	Air Force's Personnel Center
AFRL	Air Force Research Laboratory (AFWERX, innovation branch of AFRL)
AI	artificial intelligence
CES	Cyber Excepted Service
DAF	Department of the Air Force
DAFI	Department of the Air Force Instruction
DAU	Defense Acquisition University
DCAT	Defense Competency Assessment Tool
DCIPS	Defense Civilian Intelligence Personnel System
DCPDS	Defense Civilian Personnel Data System
DE	digital engineering
DECF	Digital Competency Framework
DevOps	development and operations
DevSecOps	development, security, and operations
DoD	U.S. Department of Defense
FFRDC	federally funded research and development center
FY	fiscal year
GPT	Generative Pre-Trained Transformer
GS	General Schedule
HAF	Headquarters Air Force
HAF A5/7	Headquarters, United States Air Force
HQ PACAF	Headquarters, Pacific Air Forces
IHE	institutes of higher education
KSAs	knowledge, skills, and abilities
KSAOs	knowledge, skills, abilities, and other characteristics
Lab Demo	DoD Science and Technology Laboratory Demonstration Project
LLM	large language models
MAJCOM	major command
MBSE	model-based systems engineering

MIT	Massachusetts Institute of Technology
ML	machine learning
MOOC	massive open online course
NCR	national capital region
NDS	National Defense Strategy
NLP	natural language processing
NLx	National Labor Exchange
O*NET	Occupational Information Network
OPM	U.S. Office of Personnel Management
PACAF	Pacific Air Forces
PAF	Project AIR FORCE
PCIP	Premier College Intern Program
PD	position description
PEO Digital	Program Executive Office Digital and Enterprise Services
RAI	Responsible Artificial Intelligence
RF	radio frequency
SAF/SA	Secretariat of the Air Force Studies and Analysis
SCPD	standard core personnel document
SERC	Systems Engineering Research Center
SMART	Science, Mathematics, and Research for Transformation
SME	subject-matter expert
STEM	science, technology, engineering, and mathematics
SysML	Systems Modeling Language
USAF	U.S. Air Force
USINDOPACOM	U.S. Indo-Pacific Command
USSF	U.S. Space Force

References

AFIT—*See* Air Force Institute of Technology.

AFRL—*See* Air Force Research Laboratory.

AFWERX, *2023 Annual Report*, 2023.

Air Force Biography, "Lieutenant General S. Clinton Hinote," October 2023.

Air Force Civilian Service, "Science & Engineering," webpage, undated. As of September 10, 2024:
https://afciviliancareers.com/PCIP-scienceengineering/

Air Force Futures Requirements Oversight Team, "Capability Development Overview and Operational Capability Requirements Governance," AF/A5/7 Capability Development Guidebook Volume 2A, April 11, 2022.

Air Force Institute of Technology, "By the Numbers FY 22," fact sheet, 2022.

Air Force Institute of Technology, "Digital Innovation & Integration Center of Excellence," webpage, undated. As of August 12, 2024:
https://www.afit.edu/DIICE/

Air Force Institute of Technology, Graduate School of Engineering and Management, *Academic Catalog 2023–2024*, undated.

Air Force Life Cycle Management Center, "Digital Directorate—About Us," webpage, undated. As of September 4, 2024:
https://www.aflcmc.af.mil/WELCOME/Organizations/Digital-Directorate/

Air Force Research Laboratory, "AFRL Regional Network Overview," webpage, undated-a. As of August 10, 2024:
https://afresearchlab.com/lablife/afrl-regional-network-overview/

Air Force Research Laboratory, "AFRL Scholars," webpage undated-b. As of September 11, 2024:
https://afrlscholars.usra.edu/

Air Force Research Laboratory, "Internships and Scholarships," webpage, undated-c. As of August 9, 2024:
https://afresearchlab.com/careers-and-opportunities/students-and-faculty/#intern-box

Airforce Technology, "JSTARS—Joint Surveillance and Target Attack Radar System," webpage, May 1, 2020. As of September 11, 2024:
https://www.airforce-technology.com/projects/jstars/

Air University, "Information and Facts 2022 & 2023," fact sheet, 2023.

Barno, David, and Nora Bensahel, "Addressing the U.S. Military Recruiting Crisis," *War on the Rocks*, March 10, 2023.

Berkeley ExecEd, "Data Science for Leaders Program," website, undated. As of August 28, 2024:
https://executive.berkeley.edu/programs/data-science-leaders-program

Braswell, Jasmine, "Det 4 Provides a Data Analysis Capability for ACC Units," Air Combat Command Public Affairs, October 12, 2023.

Carnegie Mellon University, School of Computer Science, Executive & Professional Education, "Artificial Intelligence," website, undated. As of August 28, 2024:
https://execonline.cs.cmu.edu/introduction-artificial-intelligence

Casserly, Jessica, "First Cohort Graduates Digital Engineering Program, Gains Vital Skills," press release, 66th Air Base Group Public Affairs, January 24, 2024.

Casserly, Jessica, "Partnership Delivers Specialized Training for Hanscom Engineers," 66th Air Base Group Public Affairs, February 28, 2022.

Castrejon, Aleah M., "AFRL Team Works to Boost Number of Advanced STEM Degrees," August 15, 2022.

Chandra, Kavitha, Sara Kraemer, Emi Aoki, Flora Stecie Norceide, and Ola Batarseh, "Integrating Model-Based Systems Engineering and Systems Thinking Skills in Engineering Courses," paper presented at 2024 ASEE Annual Conference and Exposition, June 2024.

Coe, Jessie, Maria C. Lytell, Christina Panis, and William Shelton, *Demographic Diversity of the Science, Technology, Engineering, and Mathematics (STEM) Workforce in the U.S. Department of Defense: Analysis of Compensation and Employment Outcomes*, RAND Corporation, RR-A1480-1, 2023. As of November 25, 2024:
https://www.rand.org/pubs/research_reports/RRA1480-1.html

Columbia University Data Science Institute, "Education," website, undated. As of August 28, 2024:
https://datascience.columbia.edu/education/

Coursera, "Digital Manufacturing & Design Technology Specialization," webpage, undated-a. As of September 12, 2024:
https://www.coursera.org/specializations/digital-manufacturing-design-technology

Coursera, "Introduction to Model-Based Systems Engineering," webpage, undated-b. As of September 12, 2024:
https://www.coursera.org/learn/introduction-mbse

Crider, Kim, "Air Force Data Strategy," Headquarters U.S. Air Force, briefing, undated.

DAF—*See* Department of the Air Force.

DAFI—*See* Department of the Air Force Instruction.

DAU—*See* Defense Acquisition University.

Defense Acquisition University, "CENG 003 AI Foundations for the DoD Credential," website, August 12, 2024. As of August 28, 2024:
https://icatalog.dau.edu/onlinecatalog/CredentialConceptCard.aspx?crs_id=80

Defense Acquisition University, "Course Catalog," webpage, undated-a. As of September 10, 2024:
https://www.dau.edu/courses

Defense Acquisition University, "Courses and Schedules," website, undated-b. As of August 28, 2024:
https://icatalog.dau.edu/onlinecatalog/tabnavlas.aspx

Defense Acquisition University, "DAU Glossary of Defense Acquisition Acronyms and Terms," webpage, undated-c. As of September 10, 2024:
https://www.dau.edu/glossary

Defense Acquisition University, "Digital Acquisition Modeling Workshop," course catalog, undated-d. As of September 12, 2024:
https://www.dau.edu/courses/wse-032

Defense Acquisition University, "Digital Twins for Predictive Maintenance Fundamentals," course catalog, undated-e. As of September 12, 2024:
https://www.dau.edu/courses/log-0610

Defense Acquisition University, "Models, Simulations, and Digital Engineering," course catalog, undated-f. As of September 12, 2024:
https://www.dau.edu/courses/cle-084

Defense Business Board, *Creating a Digital Ecosystem*, Business Transformation Advisory Subcommittee, DBB FY24-03, February 29, 2024.

Defense Civilian Personnel Advisory Services, "Competency Management," webpage, undated. As of September 5, 2024:
https://www.dcpas.osd.mil/policy/strategicplanning/competencymanagement

Department of the Air Force, *The United States Air Force Artificial Intelligence Annex to the Department of Defense Artificial Intelligence Strategy*, 2019.

Department of the Air Force AI Accelerator, home page, undated-a. As of August 28, 2024:
https://www.aiaccelerator.af.mil

Department of the Air Force AI Accelerator, "Education," website, undated-b. As of August 28: 2024:
https://www.aiaccelerator.af.mil/Research/Education/

Department of the Air Force AI Accelerator Public Affairs, "DAF-MIT AI Accelerator Tackles Challenge of Cultivating, Growing World-class AI Workforce," October 31, 2022.

Department of Defense Instruction 5000.97, *Digital Engineering*, Office of the Under Secretary of Defense for Research and Engineering, December 21, 2023.

Department of Defense Joint Artificial Intelligence Center and Department of Defense Chief Information Officer, *2020 Department of Defense Artificial Intelligence Education Strategy*, September 2020.

Department of the Air Force Instruction 13-504, *Nuclear Mission Professional Development*, Secretary of the Air Force, November 23, 2021.

Department of the Air Force Instruction 36-1401, *Civilian Position Classification*, Secretary of the Air Force, May 22, 2023.

Department of the Air Force Manual 36-142, *Civilian Career Field Management and Centrally Managed Programs*, Secretary of the Air Force, October 4, 2022.

Department of the Navy, *United States Navy and Marine Corps Digital Systems Engineering Transformation Strategy*, 2020.

Deputy Assistant of the Air Force (Science, Technology and Engineering), *Civilian Career Development Guide for DAF Scientist & Engineer Career Field Scientists, Engineers & Technicians*, Version 1.0, June 12, 2023.

"Development, Deployment, and Evaluation of Instructional Modules for Current and Future Practitioners of Model-Based Systems Engineering," webpage, undated. As of September 10, 2024: https://par.nsf.gov/servlets/purl/10208799

Digital Engineering Body of Knowledge, homepage, undated. As of September 10, 2024: https://de-bok.org/

Digital Transformation Office, "Learn How to Drive Successful Online Collaboration with the DAF Digital Facilitators Academy," webpage, November 20, 2023. As of August 13, 2024: https://dafdto.com/featured/facilitators-academy/

DoD—*See* U.S. Department of Defense.

Duncan, Dustin, "ICAMS Receives $9.2M to Further Model-Based Systems Engineering," Auburn University Samuel Ginn College of Engineering, March 13, 2024.

Edwards, Kathryn A., Maria McCollester, Brian Phillips, Hannah Acheson-Field, Isabel Leamon, Noah Johnson, and Maria C. Lytell, *Compensation and Benefits for Science, Technology, Engineering, and Mathematics (STEM) Workers: A Comparison of the Federal Government and the Private Sector*, RAND Corporation, RR-4267-OSD, 2021. As of September 6, 2024: https://www.rand.org/pubs/research_reports/RR4267.html

Fast Track Action Subcommittee on Critical and Emerging Technologies, *Critical and Emerging Technologies List Update*, National Science and Technology Council, February 2022.

Gehlhaus, Diana, Maria C. Lytell, James Ryseff, and Kirsten M. Keller, *Keeping Up with the Joneses: How Can DoD Address Its Technical Talent Shortage?* RAND Corporation, PT-A2884-3, 2023. As of December 7, 2023: https://www.rand.org/pubs/presentations/PTA2884-3.html

Georgia Tech, Professional Education, "Radar Systems Engineering," webpage, undated. As of September 11, 2024: https://pe.gatech.edu/courses/radar-systems-engineering

The Griffiss Institute, homepage, undated. As of September 11, 2024: https://www.griffissinstitute.org/

Groeber, Ginger, Kirsten M. Keller, Philip Armour, Samantha E. DiNicola, Irina A. Chindea, Brandon Crosby, Ellen E. Tunstall, and Shreyas Bharadwaj, *Department of the Air Force Civilian Compensation and Benefits: How Five Mission Critical and Hard-to-Fill Occupations Compare to the Private Sector and Key Federal Agencies*, RAND Corporation, RR-A334-1, 2021. As of December 7, 2023: https://www.rand.org/pubs/research_reports/RRA334-1.html

Groeber, Ginger, Paul W. Mayberry, Brandon Crosby, Mark Doboga, Samantha E. DiNicola, Caitlin Lee, and Ellen E. Tunstall, *Federal Civilian Workforce Hiring, Recruitment, and Related Compensation Practices for the Twenty-First Century: Review of Federal HR Demonstration Projects and Alternative Personnel Systems to Identify Best Practices and Lessons Learned*, RAND Corporation, RR-3168-OSD, 2020. As of December 7, 2023: https://www.rand.org/pubs/research_reports/RR3168.html

Harrington, Lisa M., Lindsay Daugherty, Craig Moore, and Tara L. Terry, *Air Force–Wide Needs for Science, Technology, Engineering, and Mathematics (STEM) Academic Degrees*, RAND Corporation, RR-659-AF, 2014. As of November 28, 2023:
https://www.rand.org/pubs/research_reports/RR659.html

Hayden, Laura, "Building Digital Literacy through Digital University," Air Combat Command Public Affairs, November 20, 2020.

Henley, Deb, "84th RADEES Optimizes Nation's LRR Systems for Air Surveillance, National Defense," 505th Command and Control Wing Public Affairs, September 23, 2021.

Hurst, J. Kyle, Steven A. Turek, Chadwick M. Steipp, and Duke Z. Richardson, *An Accelerated Future State*, Air Force Materiel Command, 2023.

Hutchinson, Nicole, Kara Pepe, Mark Blackburn, Hoon Yan See Tao, Dinesh Verma, Cliff Whitcomb, Rabia Khan, Russell Peak, and Adam Baker, "WRT-1006 Technical Report: Developing the Digital Engineering Competency Framework (DECF)—Phase 2," Stevens Institute of Technology, Systems Engineering Research Center, SERC-2021-TR-005, March 23, 2021.

Indeed Editorial Team, "Technician vs. Technologist," June 27, 2024.

Judson, Jen, "US Army Moves Out on Digital Engineering Strategy," *Defense News*, June 19, 2024.

Keller, Kirsten M., Ginger Groeber, Philip Armour, Jenna W. Kramer, Shirley M. Ross, Diana Y. Myers, Hannah Acheson-Field, Samantha E. DiNicola, Shreyas Bharadwaj, and Stephanie Williamson, *Attracting and Employing Top-Tier Civilian Technical Talent in the Department of the Air Force: A Comparison of Six Occupations with Other Federal Agencies and the Private Sector*, RAND Corporation, RR-A987-1, 2023. As of December 7, 2023:
https://www.rand.org/pubs/research_reports/RRA987-1.html

Keller, Kirsten M., Maria C. Lytell, and Shreyas Bharadwaj, *Personnel Needs for Department of the Air Force Digital Talent: A Case Study of Software Factories*, RAND Corporation, RR-A550-1, 2022. As of September 8, 2024:
https://www.rand.org/pubs/research_reports/RRA550-1.html

Keller, Kirsten M., Maria C. Lytell, David Schulker, Kimberly Curry Hall, Louis T. Mariano, John S. Crown, Miriam Matthews, Brandon Crosby, Lisa Saum-Manning, Douglas Yeung, Leslie Adrienne Payne, Felix Knutson, and Leann Caudill, *Advancement and Retention Barriers in the U.S. Air Force Civilian White Collar Workforce: Implications for Demographic Diversity*, RAND Corporation, RR-2643-AF, 2020. As of December 7, 2023:
https://www.rand.org/pubs/research_reports/RR2643.html

Knapp, David, Sina Beaghley, Troy D. Smith, Molly F. McIntosh, Karen Schwindt, Norah Griffin, Daniel Schwam, and Hanna Hoover, *DoD Cyber Excepted Service Labor Market Analysis and Options for Use of Compensation Flexibilities*, RAND Corporation, RR-A730-1, 2021. As of September 5, 2024:
https://www.rand.org/pubs/research_reports/RRA730-1.html

Lewis, Jennifer Lamping, Laura Werber, Cameron Wright, Irina Elena Danescu, Jessica Hwang, and Lindsay Daugherty, *2016 Assessment of the Civilian Acquisition Workforce Personnel Demonstration Project*, RAND Corporation, RR-1783-OSD, 2017. As of November 25, 2024:
https://www.rand.org/pubs/research_reports/RR1783.html

Libicki, Martin C., David Senty, and Julia Pollak, *Hackers Wanted: An Examination of the Cybersecurity Labor Market*, RAND Corporation, RR-430, 2014. As of September 14, 2024:
https://www.rand.org/pubs/research_reports/RR430.html

LinkedIn, *2018 Workplace Learning Report: The Rise and Responsibility of Talent Development in the New Labor Market*, 2018.

Massachusetts Institute of Technology Lincoln Laboratory, "Radar: Graduate Level—Online Course," webpage, undated-a. As of September 11, 2024:
https://www.ll.mit.edu/outreach/radar-graduate-level-online-course

Massachusetts Institute of Technology Lincoln Laboratory, "Radar: Introduction to Radar Systems—Online Course," webpage, undated-b. As of September 11, 2024:
https://www.ll.mit.edu/outreach/radar-introduction-radar-systems-online-course

Miller, Caroline M., "Military Department Personnel Posture Hearing," testimony submitted to the Subcommittee on Military Personnel, Committee on Armed Services, United States House of Representatives: Military Department Personnel Programs, March 9, 2023.

Missouri University of Science and Technology, "Systems Engineering," course catalog, undated.

MIT Professional Education, "Applied Data Science Program: Leveraging AI for Effective Decision-Making," website, undated. As of August 28, 2024:
https://professional-education-gl.mit.edu/mit-applied-data-science-course

Montgomery, Cassie, "DOD Invests $9.9M in Launch of Systems Engineering Technology Program," Auburn University Samuel Ginn College of Engineering, September 17, 2020.

National Center for Education Statistics, "The Classification of Instructional Programs," webpage, undated. As of September 13, 2024:
https://nces.ed.gov/ipeds/cipcode/

National Research Council of the National Academies, *Examination of the U.S. Air Force's Science, Technology, Engineering, and Mathematics (STEM) Workforce Needs in the Future and Its Strategy to Meet Those Needs*, 2010.

Naval Postgraduate School, "Academic Catalog," website, undated-a. As of August 28, 2024:
https://nps.smartcatalogiq.com/en/current/academic-catalog/

Naval Postgraduate School, "Systems Engineering," webpage, undated-b. As of September 12, 2024:
https://online.nps.edu/web/online/-/311-systems-engineering

Nelson, Carla, "Recent ISE Doctoral Graduate Accepts Faculty Position," Auburn University Samuel Ginn College of Engineering, September 7, 2023.

Office of the Under Secretary of Defense for Research and Engineering, "Federally Funded Research and Development Centers (FFRDC) and University Affiliated Research Centers (UARC)," webpage, undated. As of August 10, 2024:
https://rt.cto.mil/ffrdc-uarc/

O*NET Resource Center, "O*NET® 28.3 Database," webpage, undated. As of August 21, 2024:
https://www.onetcenter.org/shared/cite_resp?u=https%3A%2F%2Fwww.onetcenter.org%2Fdatabase.html%3Fref%3D404media.co&t=O%2ANET%2028.3%20Database

OPM—*See* U.S. Office of Personnel Management.

Possehl, Stephanie, and Philomena Zimmerman, "Digital Engineering and Modeling and Simulation," U.S. Department of Defense, Office of the Under Secretary of Defense, Research & Engineering, January 7, 2022.

Purdue University, The Robert H. Buckman College of Engineering Online Education Program, "Model-Based Systems Engineering Foundations and Applications to the Production Enterprise," webpage, undated. As of September 10, 2024:
https://engineering.purdue.edu/online/digital-badges/model-based-systems-engineering-foundations

"RAND Research on Civilian Workforce Issues," webpage, May 14, 2024. As of September 5, 2024:
https://www.rand.org/pubs/corporate_pubs/CPA2660-6.html

Ross, Shirley M., Rebecca Herman, Irina A. Chindea, Samantha E. DiNicola, and Amy Grace Donohue, *Optimizing the Contributions of Air Force Civilian STEM Workforce*, RAND Corporation, RR-4234-AF, 2020. As of December 7, 2023:
https://www.rand.org/pubs/research_reports/RR4234.html

Secretary of the Air Force Public Affairs, "Air Force, Space Force Announce Sweeping Changes to Maintain Superiority Amid Great Power Competition," February 12, 2024.

Shyu, Heidi, "USD(R&E) Technology Vision for an Era of Competition," Under Secretary of Defense for Research and Engineering, memorandum, February 1, 2022.

Simms, Thomas, "Engineering and Technical Management (ETM) Functional Area Framework Brief," Acting Director, Engineering Policy and Systems, Office of the Under Secretary of Defense for Research and Engineering, March 8, 2022.

Singam, Caitlyn, and Jeffrey Carter, "Model-Based Systems Engineering (MBSE)," Guide to the Systems Engineering Body of Knowledge, International Council on Systems Engineering, May 6, 2024.

Special Assistant to the Air Force Chief Scientist, "Management Initiative MI #9 Update," May 4, 2022.

Spirtas, Michael, Yool Kim, Frank Camm, Shirley M. Ross, Debra Knopman, Forrest E. Morgan, Sebastian Joon Bae, M. Scott Bond, John S. Crown, and Elaine Simmons, *Creating a Separate Space Force: Challenges and Opportunities for an Effective, Efficient, Independent Space Service*, RAND Corporation, RB-10103-AF, 2020. As of December 7, 2023:
https://www.rand.org/pubs/research_briefs/RB10103.html

Stanford University, Human-Centered Artificial Intelligence, "Professional Education," website, undated. As of August 28, 2024:
https://hai.stanford.edu/education/professional-education

Strategic Ohio Council for Higher Education, homepage, undated. As of September 13, 2024:
https://www.soche.org/

Under Secretary for Research and Engineering, "Organizational Highlight: Digital Engineering, Modeling and Simulation," webpage, February 2024. As of September 10, 2024:
https://www.cto.mil/wp-content/uploads/2024/05/Info-DEMS-2024.pdf

U.S. Air Force, "AFIT Launches Data Analysis Certificate," website, July 15, 2020. As of September 16, 2024:
https://www.wpafb.af.mil/News/Article-Display/Article/2274739/afit-launches-data-analytics-certificate-program/

U.S. Code, Title 5, Section 5101, Purpose.

U.S. Code, Title 10, Armed Forces; Subtitle A, General Military Law; Part II, Personnel, Chapter 87, Defense Acquisition Workforce; Subchapter IV, Education and Training.

U.S. Department of Defense, "Fact Sheet: 2022 National Defense Strategy," undated.

U.S. Department of Defense, *Summary of the 2018 Department of Defense Artificial Intelligence Strategy: Harnessing AI to Advance Our Security and Prosperity*, 2018.

U.S. Department of Defense, *Digital Engineering Strategy*, Office of the Deputy Assistant Secretary of Defense for Systems Engineering, June 2018.

U.S. Department of Defense, *DoD Data Strategy*, September 30, 2020.

U.S. Department of Defense, *U.S. Department of Defense Responsible Artificial Intelligence Strategy and Implementation Pathway*, DoD Responsible AI Working Council, June 2022.

U.S. Department of Defense, *DoD Strategic Management Plan: Fiscal Years 2022–2026*, July 2022.

U.S. Department of Defense, *National Defense Science & Technology Strategy 2023*, May 9, 2023.

U.S. Department of Defense, *Department of Defense Data, Analytics, and Artificial Intelligence Adoption Strategy: Accelerating Decision Advantage*, June 27, 2023.

U.S. Department of Homeland Security, "DHS STEM Designated Degree Program List," webpage, undated. As of September 13, 2024:
https://www.ice.gov/sites/default/files/documents/stem-list.pdf

U.S. Office of Personnel Management, "Competencies," webpage, undated-a. As of August 21, 2024:
https://www.opm.gov/policy-data-oversight/assessment-and-selection/competencies/

U.S. Office of Personnel Management, "Competitive Hiring," webpage, undated-b. As of September 10, 2024:
https://www.opm.gov/policy-data-oversight/hiring-information/competitive-hiring/

U.S. Office of Personnel Management, "General Schedule Qualification Policies: General Schedule Operating Manual," webpage, May 2022. As of September 25, 2024:
https://www.opm.gov/policy-data-oversight/classification-qualifications/general-schedule-qualification-policies/#url=e6

U.S. Office of Personnel Management, "Talent Management," webpage, undated-c. As of November 4, 2024:
https://www.opm.gov/policy-data-oversight/human-capital-framework/talent-management/

U.S. Office of Personnel Management, *Handbook of Occupational Groups and Families*, December 2018.

Werber, Laura, *Talent Management for U.S. Department of Defense Knowledge Workers: What Does RAND Corporation Research Tell Us?*, RAND Corporation, RR-A950-1, 2021. As of September 8, 2024:
https://www.rand.org/pubs/research_reports/RRA950-1.html